Springer Theses

Recognizing Outstanding Ph.D. Research

For further volumes:
http://www.springer.com/series/8790

Aims and Scope

The series "Springer Theses" brings together a selection of the very best Ph.D. theses from around the world and across the physical sciences. Nominated and endorsed by two recognized specialists, each published volume has been selected for its scientific excellence and the high impact of its contents for the pertinent field of research. For greater accessibility to non-specialists, the published versions include an extended introduction, as well as a foreword by the student's supervisor explaining the special relevance of the work for the field. As a whole, the series will provide a valuable resource both for newcomers to the research fields described, and for other scientists seeking detailed background information on special questions. Finally, it provides an accredited documentation of the valuable contributions made by today's younger generation of scientists.

Theses are accepted into the series by invited nomination only and must fulfill all of the following criteria

- They must be written in good English.
- The topic should fall within the confines of Chemistry, Physics, Earth Sciences, Engineering and related interdisciplinary fields such as Materials, Nanoscience, Chemical Engineering, Complex Systems and Biophysics.
- The work reported in the thesis must represent a significant scientific advance.
- If the thesis includes previously published material, permission to reproduce this must be gained from the respective copyright holder.
- They must have been examined and passed during the 12 months prior to nomination.
- Each thesis should include a foreword by the supervisor outlining the significance of its content.
- The theses should have a clearly defined structure including an introduction accessible to scientists not expert in that particular field.

Paolo Gandini

Observation of *CP* Violation in $B^{\pm} \rightarrow DK^{\pm}$ Decays

Doctoral Thesis accepted by
the University of Oxford, UK

Springer

Author
Dr. Paolo Gandini
Particle Physics
University of Oxford
Oxford
UK

Supervisor
Dr. Malcolm John
Particle Physics
University of Oxford
Oxford
UK

ISSN 2190-5053 ISSN 2190-5061 (electronic)
ISBN 978-3-319-01028-1 ISBN 978-3-319-01029-8 (eBook)
DOI 10.1007/978-3-319-01029-8
Springer Cham Heidelberg New York Dordrecht London

Library of Congress Control Number: 2013941500

Printed on acid-free paper

Springer is part of Springer Science+Business Media (www.springer.com)

"Misura ciò che è misurabile e rendi misurabile ciò che non lo è"

"Measure what is measurable, and make measurable what is not so"

Galileo Galilei

Supervisor's Foreword

The charge operator (C) swaps the electric charge of fundamental particles and the parity operator (P) reverses their direction. It would seem reasonable to assume that the physical properties of particles will not change with the application of either of these operators, and certainly not with both. How long a particle lives and the manner in which it decays should not depend on the charge, nor on its direction compared to, say, its own axis of spinning rotation.

Were you to believe these simple symmetries in the middle of last century, you would be in esteemed company. However, in an series of experimental revelations, so famous that they can be found in any particle physics textbook, first P and then CP symmetry were found to be violated under the weak nuclear force.

In 1957, a team of American physicists (Wu et al.) found that a weak decay violated parity. In a magnet, they aligned the spin angular momentum of some very cold, radioactive cobalt nuclei and then observed that, as they decayed, the ejected electron was not emitted isotropically. The weak decay, somehow stipulated a particular direction, preferentially close to the direction of the spin vector. "So what?" you might say, but this absolute directional bias meant that the mirror image of the process does not occur. In direct contrast the other fundamental forces of nature, weak decays maximally violated P symmetry.

Just as the physics community digested this result, partly by reassuring themselves that at least CP is unviolated, a clear signature of CP violation was published. In 1964 another American team (Cronin et al.) observed that long-lived "strange" mesons, assumed to consist equally of K^0 and $\bar{K}^0$ with a value of $CP = -1$, decayed 0.2 % of the time into two pions which are definitely $CP = +1$! This demonstrated that, as the neutral kaons travelled though space, the quantum-mechanical proportion of K^0 and $\bar{K}^0$ slightly altered such that a small proportion of $CP = +1$ arose. The altering of K^0 and $\bar{K}^0$ proportions are possible only if their interactions by the weak force differ. This is why the term "CP violation" immediately implies a difference between matter and antimatter (K^0 and $\bar{K}^0$ in this case).

The cosmological relevance of a matter/antimatter asymmetry in particle interactions is clear. In an inflationary model of the Big Bang, equal particle production of matter and antimatter seems logical. However at very, very early time a difference in behaviour must have been present to leave us with a Universe made exclusively of matter. In this context, the 1964 experiment would seem to have astonishing significance, but a quick calculation shows that the amount of *CP* violation in kaons is orders of magnitude too small to explain the universe. But it is a start!

Over the following decade the Standard Model of particle physics was developed around three generation of quarks, three generations of leptons and three forces (gravity excluded). In a beautifully complete revelation, it transpired that *CP* violation arises naturally in the mathematics which describes quark interaction via the weak force. And this mathematics stipulated that if there are indeed three generations of quarks, then just one parameter is the origin of all *CP* violation phenomena in the Standard Model. In 1973, it not only predicted the third quark generation now known at the "top" and "bottom" quarks, but also predicted the matter/antimatter differences should be even more glaring in the decays of bottom mesons.

The validation of this prediction had to wait until 2001 when two competing experiments, Belle (Japan) and Babar (California) discovered a large difference in the time evolution of B^0 and $\bar{B}^0$ mesons. This *CP* violation is ~ 10 times greater than the effect first observed in kaons 37 years earlier; the long delay was only due to the difficulty in producing the much-heavier bottom mesons. In the first decade of the new millennium, *CP* violation was seen in a number decay modes of the neutral B mesons. The effects varied in size but all were consistent, more-or-less with the Standard Model expectation.

Nowadays we are in a regime where the Standard Model *CP* violation must be tested to high precision by cross-comparing all the known effects as well as, of course, looking for new ones. The sticky problem is that nearly all the neutral B meson decays suffer from some level of "pollution" due to the same final state being obtainable via two quark-level processes. Hence the precision test of the Standard Model becomes clouded by uncertainty due to the interference of the two paths.

Excitingly, this pollution is essentially non-existent in a special class of charged $B^{\pm}$ decays: where that meson is reconstructed in the $DK^{\pm}$ final state. The methods for the clean measurement of the Standard Model *CP* violation were developed in the 1990s, and good attempts to apply them to the Belle and Babar datasets have been made. However, the rarity of the sensitive $B^{\pm} \rightarrow DK^{\pm}$ decays turned out to need the advent of a new machine, the Large Hadron Collider (LHC) and its dedicated B-physics experiment, LHCb.

In his thesis, Paolo Gandini describes his analysis of the first two years of LHCb data where he pioneered the optimal selection of the precious B decays in question and devised new techniques for isolating the signal from dangerous backgrounds.

He describes a landmark point where, after a decade of progress, *CP* violation in $B^{\pm} \to DK^{\pm}$ decays was observed with the gold-standard criterion of *more than five sigma*. This work already serves as a useful reference in subsequent LHC data analyses which aim to go from Paolo's discovery to a precision measurement on Standard Model *CP* violation by the end of the decade.

Oxford, May 2013 Dr. Malcolm John

Preface

CP violation is introduced in the Standard Model through the Cabibbo–Kobayashi–Maskawa mechanism (CKM) which requires that all CP violating phenomena originate from the same complex phase in the CKM quark-mixing matrix. The unitary constraint on the CKM matrix results in relations on its elements which can be conveniently represented as sums of complex numbers: the most important relation for B-physics defines a triangle (Unitary Triangle) on the complex plane, whose area is proportional to the amount of CP violation introduced by the CKM mechanism. An independent measure of the sides and the angles α, β, and γ of the Unitary Triangle is therefore crucial to test the validity of the Standard Model. At present, the CKM angles α and β are measured with good (<5 %) accuracy, while the best value of γ is still affected by a large experimental uncertainty. The LHCb experiment is expected to improve considerably the error on γ due to the large datasets it expects to collect.

The LHCb detector is a single-arm spectrometer situated on the Large Hadron Collider at CERN. It is optimised to select and reconstruct the b and c hadrons produced in high energy pp collisions, in order to study CP-violating processes and rare decays with unprecedented precision.

This thesis focuses its attention on the measurement of CP violation in $B^{\pm} \to D h^{\pm}$ decays, which can be used to determine the CKM phase γ at tree level. Charged $B \to DK$ and $B \to D\pi$ decays are reconstructed with the D meson decaying into a final state accessible by both D^0 and $\bar{D}^0$ flavour eigenstates: the resulting charge asymmetries and branching fraction ratios are directly related to and can be used in its extraction. The D intermediate state is reconstructed here into two-body charged final states: $K^{\pm}\pi^{\mp}, K^{+}K^{-}, \pi^{+}\pi^{-}$ and $\pi^{\pm}K^{\mp}$.

This thesis presents the most recent version of the analysis and it uses the full 2011 LHCb dataset of 1.0 fb^{-1}. The author has been responsible for the analysis from its early stages. The analysis has evolved considerably with time, from the very first attempts to reconstruct the favoured $B \to [K\pi]_{D\pi}$ mode in Spring 2010 to the full simultaneous strategy described in this document. The key milestones of the analysis are summarised here:

- Winter 2010
  The first branching fraction and asymmetry measurements are based on the 35.7 pb^{-1} collected in 2010 by the LHCb detector. A preliminary measurement of the ratio of branching fractions $\mathcal{B}(B^{\pm} \rightarrow DK^{\pm})/\mathcal{B}(B^{\pm} \rightarrow D\pi^{\pm})$ for the most abundant $D \rightarrow K\pi$, KK modes has been presented to the 2010 winter conferences [LHCb-CONF-2011-031].
- Summer 2011
  The first attempt to observe the suppressed $B \rightarrow [\pi K]_D K$ peak is based on the 342.7 pb^{-1} collected in the first half of 2011. This result is preliminary. The suppressed $B \rightarrow D\pi$ mode is confirmed and evidence for the suppressed $B \rightarrow DK$ mode is seen with a ignificance of 4.0 [LHCb-CONF-2011-044]. A background was missed from this preliminary analysis.
- Winter 2011
  A full simultaneous approach is adopted for the full 2011 dataset. The suppressed $B \rightarrow DK$ mode is observed with a significance $>10\sigma$. This is the analysis presented in this thesis and it has been published in [Phys.Lett.B712 (2012) 203].

This thesis is divided into six chapters:

- Chapter 1 provides the theoretical framework used to describe *CP* violation in the Standard Model (SM). After a brief overview on the CKM matrix and its parameterisations, a detailed description of the tree-level determination of the CKM angle γ using $B \rightarrow DK$ decays is given. The formalism used in the rest of the thesis is derived here.
- Chapter 2 gives a short overview of the LHCb detector, the experimental apparatus used to perform the analysis described in this thesis.
- Chapter 3 describes the selection strategies developed for the analysis. Signal candidates are selected exploiting the topology of the decays of interest. As LHCb is a high background environment, the choice of a high performance selection is crucial to ensure an adequate signal efficiency and background rejection. The selection is tuned in order to maximise the significance of the suppressed $B \rightarrow DK$ peak, with the goal of its observation.
- Chapter 4 includes a detailed background description for all the signal modes considered. Background studies are performed using both simulated events and, where possible, data. Extra care is taken in describing peaking backgrounds coming from charmless modes and internal cross feed. All the PDFs used in the fit to the *CP* observables are described here.

- Chapter 5 presents a detailed description of the simultaneous maximum likelihood fit used in the extraction of the *CP* physics observables. The fitter is designed to consider all the different signal modes at the same time and take advantage of a large set of parameters in common. This chapter concludes with a detailed study of the systematic uncertainties on the measured quantities.
- Chapter 6 summarises the numerical results of the fit to data, including a comparison with the existing results and a preliminary fit to the CKM phase γ.

Abstract

An accurate determination of the angle γ of the Unitary Triangle is one of the most important goals of the LHCb experiment. The LHCb detector is a single-arm spectrometer at the LHC, optimised for beauty and charm flavour physics. As the angle γ is the least experimentally constrained parameter of the Unitary Triangle, its precise experimental determination can be used to test the validity of the Standard Model. The Unitary Triangle phase γ can be extracted in $B \to DK$ decays at tree-level, exploiting the interference between $b \to c(\bar{u}s)$ and $b \to u(\bar{c}s)$ transitions. This interference is sensitive to γ and can give measurable charge asymmetries. In particular, $\gamma \neq 0$ is required to produce *direct CP* violation in B decays and this is the only *CP*-violating mechanism for the decay of charged $B^{\pm}$ mesons.

In this thesis, an analysis of *CP* violation in $B^{\pm} \to DK^{\pm}$ and $B^{\pm} \to D\pi^{\pm}$ decays is presented, where the D meson is reconstructed in the two-body final states: $K^{\pm}\pi^{\pm}$, $K^{+}K^{-}$, $\pi^{+}\pi^{-}$ and $\pi^{\pm}K^{\pm}$. The analysis uses the full 2011 LHCb dataset of 1.0 fb^{-1}, collected from pp collisions at $\sqrt{s} = 7$ TeV. Several *CP*-related quantities, e.g. the ratio of $B \to DK$ and $B \to D\pi$ branching fractions and their charge asymmetries, are measured via a simultaneous fit to the invariant mass distributions of the modes considered. The suppressed $B \to DK^{\pm}$ mode is observed for the first time with $\approx 10\sigma$ significance. Once all measurements are combined, direct *CP* violation is established in $B^{\pm}$ decays with a total significance of 5.8σ. The measured *CP* observables are summarised here with their statistical uncertainties and assigned systematic uncertainties:

$$R_{\mathrm{CP+}} = 1.007 \pm 0.038 \pm 0.012$$
$$A_{\mathrm{CP+}} = 0.145 \pm 0.032 \pm 0.010$$
$$R_{\mathrm{ADS(K)}} = 0.0152 \pm 0.0020 \pm 0.0004$$
$$A_{\mathrm{ADS(K)}} = -0.52 \pm 0.15 \pm 0.02$$
$$R_{\mathrm{ADS}(\pi)} = 0.00410 \pm 0.00025 \pm 0.00005$$
$$A_{\mathrm{ADS}(\pi)} = 0.143 \pm 0.062 \pm 0.011$$

Contents

1 Introduction 1
1.1 Overview of the Standard Model (SM) 1
1.2 Discrete Symmetries 3
1.3 Mixing Formalism 4
1.4 The Three Types of *CP* Violation 6
1.4.1 Direct *CP* Violation 7
1.4.2 *CP* Violation in Mixing 7
1.4.3 *CP* Violation in the Interference Between Decay and Mixing 7
1.5 *CP* Violation in the Standard Model 8
1.5.1 The CKM Matrix 9
1.5.2 The CKM Parameterisations 10
1.5.3 The Unitary Triangle (UT) 11
1.5.4 Global Fits to V_{CKM} 13
1.6 Tree Level Determination of γ with $B \to DK$ Decays 14
1.6.1 GLW Method 16
1.6.2 ADS Method 17
1.6.3 Physical Observables 18
References 22

2 The LHCb Experiment at the LHC 25
2.1 The LHC at CERN 25
2.2 The LHCb Detector 28
2.2.1 LHCb Detector Running Conditions 30
2.2.2 VELO 32
2.2.3 Dipole Magnet 35
2.2.4 Tracking Stations 36
2.2.5 RICH 38
2.2.6 Calorimeters 39
2.2.7 Muon Stations 43
2.2.8 Trigger 44
2.2.9 Track Reconstruction 47

2.2.10 Particle Identification (PID). 48
2.2.11 LHCb Software Framework. 51
References 52

3 Reconstruction and Selection 55
3.1 Overview of the $B \to [hh]_D h$ Decay Modes. 55
3.2 The First Favoured $B \to [K\pi]_D \pi$ Candidate. 56
3.3 Datasets and MC Samples. 57
3.4 Stripping. 59
3.5 Selection: A MVA Approach. 60
3.5.1 Input Samples and Variables 61
3.5.2 Performance and Optimal Cut Evaluation 63
3.6 Trigger Requirements. 68
3.7 Multiple Candidate Arbitration 69
3.8 Total Efficiencies and Expected Yields. 69
3.9 Resolution and Mass Lineshapes 71
3.9.1 Correctly Reconstructed $B \to [hh]_D h$ Peaks 72
3.9.2 D Mass Constraint and MC Lineshapes 73
3.9.3 Lineshapes in Data and Signal Mass Windows 75
3.9.4 $B \to Dh$ Misidentified Shape. 78
3.10 Momentum Rescaling. 80
3.11 PID Calibration 83
References 85

4 Background Studies. 87
4.1 Combinatorial Background 87
4.2 Backgrounds from Decays with Muons. 89
4.2.1 Charmonium Veto 89
4.2.2 Semileptonic Backgrounds 91
4.3 Partially Reconstructed Backgrounds 92
4.3.1 Partially Reconstructed Topology. 93
4.3.2 Partially Reconstructed Inclusive Sample and PDFs . . . 96
4.4 Charmless Backgrounds 104
4.4.1 Selecting on D Flight Distance 105
4.4.2 Charmless Expectations from MC 106
4.4.3 Charmless Expectations from Data Sidebands: $[KK, \pi\pi]_D$ 109
4.5 $B \to [hh]_D h$ Internal Cross Feed. 112
4.5.1 Misidentification of the $[hh]_D$ Daughter Tracks 113
4.5.2 Internal Bachelor-Daughter Combinatorial Background. 114
4.5.3 Favoured → Suppressed-Mode Cross Feed 116
References 122

5 The Simultaneous Fit . . . 123
5.1 Fitter Strategy . . . 123
5.1.1 Introduction . . . 123
5.1.2 Fitter Components . . . 125
5.1.3 Yields . . . 127
5.1.4 Charge Asymmetries . . . 127
5.2 Validation . . . 129
5.3 Systematic Uncertainties . . . 130
5.3.1 Multiplicative Corrections on $R^{hh}_{K/\pi}$ . . . 131
5.3.2 Systematic Error on the Bachelor PID Variable . . . 133
5.3.3 Propagation Through the Final Fit . . . 136
References . . . 139

6 Results . . . 141
6.1 Numerical Fit Results . . . 141
6.2 Yield Comparison . . . 142
6.3 The *CP* Observables . . . 146
6.4 Statistical Significance of the Measurements . . . 148
6.5 Comparison with Existing Measurements . . . 149
6.6 A Preliminary Fit to the CKM Angle γ . . . 150
6.6.1 Likelihood Scan Including Only $B \rightarrow [hh]_D K$ Observables . . . 151
6.6.2 Likelihood Scan Including the GGSZ Analysis . . . 153
6.6.3 Including the Systematical Uncertainties on r_D and δ_D . . . 153
References . . . 155

7 Conclusions . . . 157
Reference . . . 158

Appendix A: Distributions of the BDT Input Variables . . . 159

Appendix B: Validation of the A_{p_T} Variable . . . 161

Appendix C: Histograms of Systematic Errors . . . 163

Appendix D: Curriculum Vitae . . . 167

Chapter 1
Introduction

This chapter introduces the Standard Model (SM) and the theoretical framework used to describe the *CP* violation. It starts with a brief general introduction to the SM (Sect. 1.1) and discrete symmetries (Sect. 1.2). A short description of the mixing formalism (Sect. 1.3) for neutral meson systems is also presented. Although the main focus of this work is the observation of direct *CP* violation in charged *B* mesons, the mixing formalism is useful to introduce the three possible ways to observe *CP* violation in meson decays, described in Sect. 1.4. The CKM matrix and the Unitarity Triangle (UT) are described in Sect. 1.5. The chapter concludes with a description of the tree level determination of the CKM angle γ using $B \to DK$ decays (Sect. 1.6) including the formalism relevant to the interesting physical observables used in the rest of this thesis.

1.1 Overview of the Standard Model (SM)

The Standard Model (SM) of particle physics describes the properties of all known elementary particles and their interactions [1, 2]. Born in the late sixties and tested for more than 40 years, it has been proven to be the most successful predictive theory in physics. Despite its success, the SM has limits as it does not include gravity as a fundamental force. Furthermore, no candidate for dark matter is included in the theory and the amount of *CP* violation dictis insufficient to explain the abundance of matter in our universe. The SM is a relativistic quantum field theory [3] which includes three generations of fundamental fermions (spin-1/2 particles) and their analogous antiparticles which have opposite quantum numbers:

P. Gandini, *Observation of CP Violation in* $B^{\pm} \to DK^{\pm}$ *Decays*,
Springer Theses, DOI: 10.1007/978-3-319-01029-8_1,

Quarks

$$\begin{bmatrix} u \\ d \end{bmatrix} \begin{bmatrix} c \\ s \end{bmatrix} \begin{bmatrix} t \\ b \end{bmatrix}$$

Leptons

$$\begin{bmatrix} \nu_e \\ e \end{bmatrix} \begin{bmatrix} \nu_\mu \\ \mu \end{bmatrix} \begin{bmatrix} \nu_\tau \\ \tau \end{bmatrix}$$

The three generations have similar properties, but the particles involved have different masses. The SM includes three forces by which fermions may interact:

Strong	Electromagnetic	Weak
g 8 gluons	γ 1 photon	$W^{\pm}, Z^0$ 3 weak bosons

Forces are mediated by bosons (integer-spin particles): the photon is the qantum of the electromagnetic field, the $W^{\pm}$ and Z^0 bosons mediate the weak force and eight flavours of gluon carry the strong force. The charged leptons interact via the electromagnetic and weak forces and neutrinos feel only the weak force. Quarks couple to all the three forces.

As quarks can interact strongly, they are never isolated in nature but always bound in composite subatomic particles called *hadrons*. *Baryons* are made up of three quarks and hence have half-integer spin, while integer-spin *mesons* are formed by a quark and antiquark pair. Most relevant to this thesis are B mesons, which are composed of a b quark and other lighter quarks, e.g. $B^-(b\bar{u})$, $\bar{B}^0(b\bar{d})$ and $\bar{B}^0_s(b\bar{s})$. In order to understand the properties of the b quark, B mesons, whose behaviour is dominated by the heavier b quark, are an excellent particle to study.

A core part of the SM is the concept of symmetry. The laws of physics are invariant, or symmetric, under simple transformations of the coordinate system like a rotation. Analogously, symmetry in a Lagrangian represents some fundamental structure of the theory. For example, the behaviour of the eight gluons derives from the symmetry of the underlying Lagrangian. The part of mathematics which deals with symmetries is *group theory* with which the SM can be summarised with the famous structure:

$$\mathrm{SU}(3)_C \times \mathrm{SU}(2)_L \times \mathrm{U}(1)_Y.$$

The first term refers to the theory of the strong interaction (Quantum ChromoDynamics, QCD [4]), while the second and third parts describe the unified electroweak force. Each term represents a beautiful internal symmetry but which does not completely describe all the particle phenomena. The perfect symmetry described by the electroweak part is spontaneously broken at low energy via the Higgs mechanism [5, 6], which gives masses to the weak bosons $W^{\pm}$ and Z^0 while keeping the electromagnetic photon massless. As a result, a new extra scalar (spin-0) particle appears in the model, usually referred to as the Higgs boson. The interaction with the Higgs field is also used to give masses to the known fermions without violating the gauge invariance of the theory and gives rise to quark mixing, as will be explained in Sect. 1.5.1.

Violations of the expected symmetries of a theory play a crucial role in the understanding of particle physics. Most relevant to this thesis is the violation of the C and P symmetries, which will be described shortly. It turns out that B mesons are an excellent place to observe this phenomenon as the SM predicts large effects with high accuracy. CP violation has cosmological significance as it plays a crucial role in baryogenesis. It is one of the Sakharov conditions [7][1]:

- Violation of the baryon number,
- C and CP violation,
- Interactions outside of thermal equilibrium,

which set out the conditions necessary to produce an abundance of matter in the universe.

1.2 Discrete Symmetries

Symmetry is an important concept in any theory that describes a physical system since it results in conserved quantities, as stated by the Noether's theorem [8]. In the Standard Model (SM), three discrete transformations can be defined:

- **Parity (*P*)**
  The parity operator reverses the spatial axes, $(t, \vec{x}) \rightarrow (t, -\vec{x})$. It flips the momenta of particles keeping their spin unchanged. Parity is implemented by a unitary operator $U(P)$ such as $U(P)^2 = I$, which requires its eigenstates to be equal to ± 1.
- **Charge conjugation (*C*)**
  Charge conjugation changes particles with their corresponding antiparticles. Both momentum and energy components are unchanged.
- **Time reversal (*T*)**
  Time reversal exchanges the future and past light-cones, inverting the time axis only $(t, \vec{x}) \rightarrow (-t, \vec{x})$.

The assumptions of Lorentz invariance and locality guarantee that each phenomenon is CPT invariant [9]. Furthermore any interacting theory that violates this CPT invariance necessarily violates Lorentz invariance [10]. However, no requirement is made on the single C, P and T symmetries, which can be individually violated. P was found to be maximally violated in weak decays by C. S. Wu et al. in 1957

[1] Although these conditions have been attributed to Sakharov, they were not listed explicitly in his paper in the current form.

[11], while *CP* was believed to be preserved in all interactions until 1964, when a small ($\mathcal{O}(10^{-3})$) *CP* violation was discovered in neutral kaon decays by J. Cronin and V. Fitch [12]. *CP* violation is now a well established phenomenon, also confirmed and observed in neutral *B* meson decays, with results from the *b*-factories (BaBar [13] and Belle [14]) and the Tevatron (CDF [15] and D0 [16]). *T*-violation has also been directly observed in the neutral-kaon system [17].

1.3 Mixing Formalism

CP violation can manifest itself in different ways both in charged and neutral meson decays. In order to present all the observable types of *CP* violation, it is important to introduce the concept of meson mixing. This occurs only in the neutral meson systems (K^0, D^0, B^0, B_s^0), where particles and antiparticles can oscillate before decaying, and mixing processes play an important role in *CP*-violating phenomena. In charged mesons, instead, charge conservation forbids oscillation and *CP* violation is possible only when the decay matrix elements are different (Sect. 1.4).

Since X^0 and $\bar{X}^0$ states can mix via the box diagrams shown in Fig. 1.1, the mass eigenstates of the system, defined as the eigenstates of the Hamiltonian, can differ from the flavour eigenstates, which are defined by their $q\bar{q}$ quark content. If *CP* were an exact symmetry, the mass eigenstates would be also exact *CP* eigenstates; when this is not the case, states can mix with each other propagating in time according to the dynamics of the two-state system. If $|X^0\rangle$ represents a generic flavour eigenstate, the mixed state can be written as the superposition:

$$|\psi(t)\rangle = a(t)|X^0\rangle + b(t)|\bar{X}^0\rangle \tag{1.1}$$

which evolves according to the time dependent Schrödinger equation:

$$i\frac{\partial}{\partial t}\begin{bmatrix} a(t) \\ b(t) \end{bmatrix} = \mathcal{H}\begin{bmatrix} a(t) \\ b(t) \end{bmatrix} \tag{1.2}$$

where $\mathcal{H}$, the Hamiltonian of the system may be:

$$\mathcal{H} = M - i\frac{\Gamma}{2} \equiv \begin{pmatrix} m_{11} & m_{12} \\ m_{12}^* & m_{22} \end{pmatrix} - \frac{i}{2}\begin{pmatrix} \gamma_{11} & \gamma_{12} \\ \gamma_{12}^* & \gamma_{22} \end{pmatrix} \tag{1.3}$$

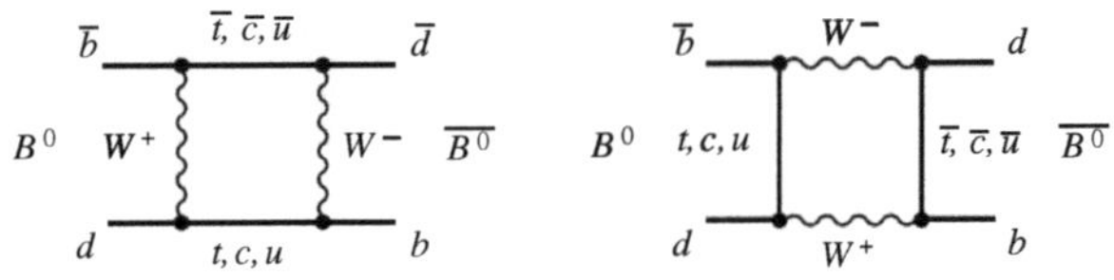

Fig. 1.1 Box diagrams for B^0–$\bar{B}^0$ mixing as an example of neutral meson mixing

$\mathcal{H}$ is the Hamiltonian of the system, decomposed into a *mass* matrix M, which describes the propagation of the states in time, and a *decay* matrix Γ. Both matrices are Hermitian and the off-diagonal elements represent the mixing terms in propagation (m_{12}) and decay (γ_{12}). In particular, γ_{12} takes into account any common decay mode of $|X^0\rangle$ and $|\bar{X}^0\rangle$. CPT invariance requires that $m_{11} = m_{22}$ and $\gamma_{11} = \gamma_{22}$.

The eigenvalues of $\mathcal{H}$, $\mu_{L,H}$, can be conveniently expressed as:

$$\mu_L = m_L - i\frac{\Gamma_L}{2} \qquad \mu_H = m_H - i\frac{\Gamma_H}{2} \tag{1.4}$$

where the mass ($m_{L,H}$) and decay parts ($\Gamma_{L,H}$) are highlighted separately. The subscripts L and H indicate the lighter and heavier eigenstates respectively. The corresponding mass eigenstates are defined as the eigenvectors of the Hamiltonian:

$$\begin{aligned} |X_L\rangle &= p|X^0\rangle + q|\bar{X}^0\rangle \\ |X_H\rangle &= p|X^0\rangle - q|\bar{X}^0\rangle \end{aligned} \tag{1.5}$$

where $|X_L\rangle$ and $|X_H\rangle$ label the lighter and heavier mass eigenstates respectively and q and p are complex numbers. The normalisation of the states imposes $|q|^2 + |p|^2 = 1$. Defining the two differences $\Delta m = m_H - m_L$ and $\Delta\Gamma = \Gamma_H - \Gamma_L$, one can obtain the following relations in term of mass and decay differences:

$$(\Delta m)^2 - \frac{1}{4}(\Delta\Gamma)^2 = 4(|m_{12}|^2 - |\gamma_{12}|^2) \quad \Delta m \Delta\Gamma = 4\mathrm{Re}(m_{12}\gamma_{12}^*) \tag{1.6}$$

$$\frac{q}{p} = \sqrt{\frac{m_{12}^* - \frac{1}{2}i\gamma_{12}^*}{m_{12} - \frac{1}{2}i\gamma_{12}}} = -\frac{\Delta m - \frac{i}{2}\Delta\Gamma}{2(m_{12} - \frac{i}{2}\gamma_{12})} \tag{1.7}$$

Δm and $\Delta\Gamma$ are good measurable observables, accessible experimentally.

Another useful basis that can be used to describe the X_0–$\bar{X}_0$ system uses the two *CP* eigenstates $|X_+\rangle$ and $|X_-\rangle$. Using the convention[2] that $CP|X^0\rangle = |\bar{X}^0\rangle$ and $CP|\bar{X}^0\rangle = |X^0\rangle$, the two $(+, -)$ orthogonal *CP* eigenstates are defined as:

$$|X_+\rangle = \frac{1}{\sqrt{2}}(|X^0\rangle + |\bar{X}^0\rangle) \qquad |X_-\rangle = \frac{1}{\sqrt{2}}(|X^0\rangle - |\bar{X}^0\rangle) \tag{1.8}$$

which relate to the (H, L) basis via:

$$|X_L\rangle = \frac{p+q}{\sqrt{2}}|X_+\rangle + \frac{p-q}{\sqrt{2}}|X_-\rangle \qquad |X_H\rangle = \frac{p+q}{\sqrt{2}}|X_-\rangle - \frac{p-q}{\sqrt{2}}|X_+\rangle \tag{1.9}$$

[2] A more general definition of the *CP* operator would be $CP|X^0\rangle = e^{+i\xi}|\bar{X}^0\rangle$ and $CP|\bar{X}^0\rangle = e^{-i\xi}|X^0\rangle$, but the phase ξ is arbitrary and is chosen to be equal to 0 here for simplicity.

Whenever the (L, H) states differ from the CP eigenstates $(+, -)$, CP violation occurs. Non-conservation of CP, initially predicted by Gell-Mann and Pais in 1955 [18], was discovered by J. Cronin and V. Fitch in $K_L^0 \to \pi^+\pi^-$ decays [12]. In fact, the K_L^0 state can be considered only approximately a CP-odd eigenstate, which would decay into a three-pion final state. However, since this symmetry is not perfect, K_L^0 is also found to decay into the CP even eigenstate $\pi^+\pi^-$ with a branching fraction of about 2×10^{-3} [1].

The same formalism applies to every X^0–$\bar{X}^0$ system, thus the Hamiltonian, and so the Δm and $\Delta\Gamma$ observables, differ. In the B^0 system, the oscillation frequency is comparable to the decay time, while for B_s^0 mesons oscillations are faster and for D^0 mesons slower. In the case, for example, of the B^0 meson, where $\Delta\Gamma \ll \Delta m$, the larger mass of the b quark allows more phase space for the decays of both B^0 and $\bar{B}^0$ and the lack of decay suppression of one of the two eigenstates implies nearly equivalent lifetimes. With the approximation $\Delta\Gamma \ll \Delta m$, the following set of equations holds:

$$|B^0(t)\rangle = f_+(t)|B^0\rangle + \frac{q}{p} f_-(t)|\bar{B}^0\rangle \tag{1.10}$$

$$|\bar{B}^0(t)\rangle = \frac{p}{q} f_+(t)|B^0\rangle + f_-(t)|\bar{B}^0\rangle \tag{1.11}$$

with

$$f_+(t) = e^{-imt}e^{-\Gamma t/2}\cos(\Delta m\, t/2) \qquad f_-(t) = e^{-imt}e^{-\Gamma t/2} i \sin(\Delta m\, t/2) \tag{1.12}$$

and $m = (m_H + m_L)/2$, $\Gamma = (\Gamma_H + \Gamma_L)/2$. It should be noted that $\Delta m/2$ is the oscillation frequency of the two flavour eigenstates.

1.4 The Three Types of *CP* Violation

Using the following generic decay amplitudes of a meson X (charged or neutral) to a final state f,

$$\begin{aligned} \langle f|\mathcal{H}|X\rangle = A_f \quad \langle f|\mathcal{H}|\bar{X}\rangle = \bar{A}_f \\ \langle \bar{f}|\mathcal{H}|X\rangle = A_{\bar{f}} \quad \langle \bar{f}|\mathcal{H}|\bar{X}\rangle = \bar{A}_{\bar{f}} \end{aligned} \tag{1.13}$$

it is possible to distinguish three distinct CP-violating effects.

1.4.1 Direct CP Violation

Direct *CP* violation refers to *CP* violation in the decay. It occurs when the two decay amplitudes A_f and $\bar{A}_{\bar{f}}$ differ in modulus, so $|A_f| \neq |\bar{A}_{\bar{f}}|$. This results in different rates of $X \rightarrow f$ with respect to its *CP* conjugate process $\bar{X} \rightarrow \bar{f}$. This phenomenon can occur both in charged and neutral mesons, it depends on the specific mode (the final state f) and it is the only *CP*-violating mechanism for the decay of charged mesons. The total $X \rightarrow f$ decay amplitudes can be generally expressed as:

$$A_f = \sum_i A_i e^{i(\phi_i + \delta_i)} \quad \text{and} \quad \bar{A}_{\bar{f}} = \eta_{CP} \sum_j A_j e^{i(-\phi_j + \delta_j)} \tag{1.14}$$

where η_{CP} is the *CP* eigenvalue if f is a *CP* eigenstate, ϕ_i are the *CP*-odd phases and δ_i any *CP*-even phase. The decay can occur via several different processes (indicated with the sum over all the amplitudes). From Eq. 1.14, one can write:

$$|A|^2 - |\bar{A}|^2 = -2 \sum_{i,j} A_i A_j \sin(\phi_i - \phi_j) \sin(\delta_i - \delta_j) \tag{1.15}$$

So, direct *CP* violation occurs only if more than one amplitude is involved in the decay and the relative weak and strong phases are different. This type of *CP* violation is responsible for all the charge asymmetries measured in the analysis presented in this thesis. The specific formalism used for $B^{\pm} \rightarrow DK^{\pm}$ and $B^{\pm} \rightarrow D\pi^{\pm}$ decays is described in detail in Sect. 1.6.

1.4.2 CP Violation in Mixing

CP violation in mixing occurs when the mass eigenstates of a neutral meson are not exact *CP* eigenstates. This type of *CP* violation is usually referred to as *indirect CP violation* and doesn't depend on the specific decay mode. From the definitions given in Eq. 1.9, the Hamiltonian eigenstates are simultaneous eigenstates of the *CP* operator only if $|q/p| = 1$. This type of *CP* violation is responsible for $K_L^0 \rightarrow \pi^+\pi^-$ decays [12].

1.4.3 CP Violation in the Interference Between Decay and Mixing

CP violation can manifest itself even if $|q/p| = 1$ and $|A_f| = |\bar{A}_{\bar{f}}|$ at the same time. This is the case when both X^0 and $\bar{X}^0$ mesons can decay into the same final state f (thus $f = \bar{f}$) and interference between the decay process and mixing occurs,

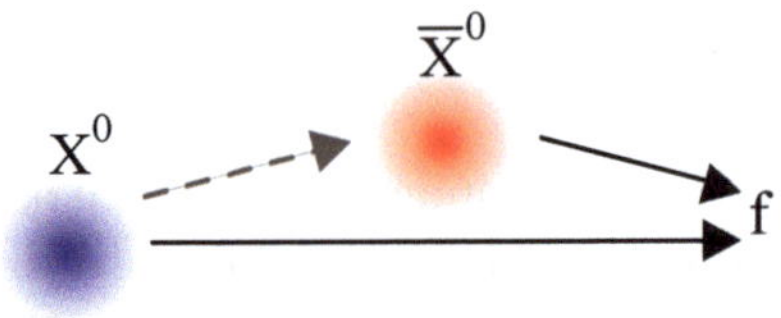

Fig. 1.2 Interference between decay and mixing in the X^0 and $\bar{X}^0$ system. This occurs only if the final state f is accessible by both X^0 and $\bar{X}^0$ states

as sketched in Fig. 1.2. In this case, the following time-dependent asymmetry can be considered:

$$A_{CP}(t) = \frac{\Gamma(X^0(t) \to f) - \Gamma(\bar{X}^0(t) \to f)}{\Gamma(X^0(t) \to f) + \Gamma(\bar{X}^0(t) \to f)} \tag{1.16}$$

By construction, Eq. 1.16 requires the tagging of the flavour of the X^0 and $\bar{X}^0$ states at the point of $X \to f$ decay. Using Eq. 1.10 and Eq. 1.11, $A_{CP}(t)$ can be rewritten as:

$$A_{CP}(t) = C\cos(\Delta m\, t) - S\sin(\Delta m\, t) \tag{1.17}$$

where

$$C = \frac{1-|\lambda|^2}{1+|\lambda|^2}, \qquad S = \frac{2\,\mathrm{Im}\,\lambda}{1+|\lambda|^2} \qquad \text{with} \qquad \lambda = \frac{q}{p}\frac{\overline{A}_f}{A_f}. \tag{1.18}$$

The oscillatory pattern of $A_{CP}(t)$ depends on the complex parameter λ and can differ from zero even if $|q/p| = 1$ and $|A_f| = |\bar{A}_{\bar{f}}|$. That is the case if λ is a pure phase and $\Im\lambda \neq 0$, leading to a non-zero S coefficient. Furthermore, unlike direct *CP* violation, no multiple decay amplitudes and relative different weak and strong phases are required. This type of *CP* violation is exploited in $B^0 \to J/\Psi K^0$ and $B^0_s \to J/\Psi\phi$ decays.

1.5 *CP* Violation in the Standard Model

CP violation in the Standard Model (SM)[3] originates from quark mixing as introduced first by M. Kobayashi and T. Maskawa in 1973 [19] with a generalisation of the mixing mechanism pioneered by N. Cabibbo in 1963 [20] and S. Glashow, J. Iliopoulos and L. Maiani in 1970 [21]. *CP* violation is explained via a complex phase in the quark mixing matrix, which can be introduced in the theory only when three (or more) families of fermions are present. The prediction of a third fermion generation preceded its discovery, and even that of the charm quark, discovered the year after [22] with the J/ψ resonance. The bottom quark was discovered later in 1977 [23].

[3] The strong *CP* problem and *CP* violation in the lepton sector are not treated here.

1.5.1 The CKM Matrix

In the SM, the mass terms for the quark fields originate from Yukawa interactions of the fermionic fields with the Higgs [1]. For a three-family theory the Yukawa Lagrangian is:

$$\mathcal{L}_Y = -Y^d_{ij}\,\overline{Q^I_{Li}}\,\phi\, d^I_{Rj} - Y^u_{ij}\,\overline{Q^I_{Li}}\,\varepsilon\,\phi^* u^I_{Rj} + \text{h.c.} \tag{1.19}$$

where ϕ is the Higgs field, ε is the 2×2 completely antisymmetric tensor, Q^I_L are the left-handed $SU_L(2)$ quark doublets and d^I_R (u^I_R) the right-handed singlets. In this weak eigenstate basis, $Y^{u,d}$ are 3×3 complex coupling matrices and the sum on i, j indicates the sum on the different quark families. When the Higgs field acquires a vacuum expectation value (v), mass terms appear in the Lagrangian described by two 3×3 mass matrices:

$$\mathcal{L}_{\text{mass}} = -M^d_{ij}\,\overline{d^I_{Li}}\, d^I_{Rj} - M^u_{ij}\,\overline{u^I_{Li}}\, u^I_{Rj} + \text{h.c.} \tag{1.20}$$

where

$$M^u = \frac{vY^u}{\sqrt{2}} \qquad M^d = \frac{vY^d}{\sqrt{2}} \tag{1.21}$$

In principle, the mass matrices $M^{u,d}$ can be non-diagonal (which leads to quark mixing) in the flavour basis and it is always possible to diagonalise them, moving to the mass eigenstate basis, via a unitarity transformation:

$$U^u_L M^u U^{u\dagger}_R = \text{diag}(m_u, m_c, m_t) \tag{1.22}$$

$$U^d_L M^d U^{d\dagger}_R = \text{diag}(m_d, m_s, m_b) \tag{1.23}$$

with the corresponding rotations on the left-handed quarks:

$$d'_{Li} = U^d_{Lij} d_{Lj} \qquad u'_{Li} = U^u_{Lij} u_{Lj} \tag{1.24}$$

This has a profound effect on charged weak interactions, causing quark mixing. In fact, weak currents can now be expressed as:

$$J^\mu_{W^+} = \frac{1}{\sqrt{2}}\bar{u}_L \gamma^\mu (U^{u\dagger}_L U^d_L) d_L \qquad J^\mu_{W^-} = \frac{1}{\sqrt{2}}\bar{d}_L (U^{d\dagger}_L U^u_L)\gamma^\mu u_L \tag{1.25}$$

with

$$U^{u\dagger}_L U^d_L = V_{\text{CKM}} = \begin{pmatrix} V_{ud} & V_{us} & V_{ub} \\ V_{cd} & V_{cs} & V_{cb} \\ V_{td} & V_{ts} & V_{tb} \end{pmatrix} \tag{1.26}$$

As $U_L^u \neq U_L^d$ in general, V_{CKM} is not diagonal and mixing of the different families of up-type and down-type quarks can occur. It should be noted that this kind of transformation doesn't affect neutral interactions. As the U matrices are unitary ($UL^{u\dagger}U_L^u = U_L^{d\dagger}U_L^d = I$):

$$\begin{aligned} \bar{u}_L\gamma^\mu(U_L^{u\dagger}U_L^u)u_L &= \bar{u}_L\gamma^\mu I u_L \\ \bar{d}_L\gamma^\mu(U_L^{d\dagger}U_L^d)d_L &= \bar{d}_L\gamma^\mu I d_L \end{aligned} \tag{1.27}$$

and neutral currents are therefore diagonal. This absence of fundamental Flavour Changing Neutral Currents in the SM is in agreement with experimental data. However, the V_{CKM} matrix is in principle not diagonal and it generates flavour changing charged currents.

V_{CKM} is a 3×3 complex unitary matrix, which can be parameterised by 3^2 real numbers. However, once the arbitrary phases of the quark fields are reabsorbed, the number of physical significant parameters reduces to four, three angles and one phase. This single complex phase is responsible for all CP-violating processes in the SM, which are thus all correlated phenomena by construction, making the model a very predictive theory on CP-related observables. A precise experimental determination of all the V_{CKM} elements, which are in principle free parameters of the model, provides strong constraints on the validity of the theory itself.

1.5.2 The CKM Parameterisations

Several representations of the V_{CKM} matrix have been proposed in literature and the most common are reported here.

The Standard Parameterisation

This choice was proposed in [24]. V_{CKM} is expressed in an exactly unitary form as the product of three rotation matrices and one complex phase:

$$V_{\text{CKM}} = \begin{pmatrix} c_{12}c_{13} & s_{12}c_{13} & s_{13}e^{-i\delta} \\ -s_{12}c_{23} - c_{12}s_{23}s_{13}e^{i\delta} & c_{12}c_{23} - s_{12}s_{23}s_{13}e^{i\delta} & s_{23}c_{13} \\ s_{12}s_{23} - c_{12}c_{23}s_{13}e^{i\delta} & -c_{12}s_{23} - s_{12}c_{23}s_{13}e^{i\delta} & c_{23}c_{13} \end{pmatrix} \tag{1.28}$$

where $c_{ij} = \cos\theta_{ij}$ and $s_{ij} = \sin\theta_{ij}$ for $i < j = 1, 2, 3$. θ_{12}, θ_{13} and θ_{23} are the three Euler angles which represent the mixing angles between the three quark generations. The complex phase δ, if non-zero, is responsible for every CP-violating phenomena in the Standard Model.

The Wolfenstein Parameterisation

The elements of the CKM matrix exhibit a hierarchical pattern as $s_{13} \ll s_{23} \ll s_{12} \ll 1$. Experimentally, the elements on the diagonal are found to be close to unity and the off-diagonal elements are closer to zero the further they are from the diagonal itself. This pattern is exploited in the *Wolfenstein parameterisation* [25] where all the elements are expanded as a power series of the parameter $\lambda = s_{12}$, which is the sinus of the Cabibbo angle $\sin\theta_C \approx 0.22$. With the following substitutions:

$$\begin{aligned} s_{12} &\equiv \lambda, \\ s_{23} &\equiv A\lambda^2 \\ s_{13}e^{-i\delta} &\equiv A\lambda^3(\rho - i\eta) \end{aligned} \tag{1.29}$$

the CKM matrix becomes:

$$V_{\mathrm{CKM}} = \begin{pmatrix} 1-\frac{\lambda^2}{2} & \lambda & A\lambda^3(\rho - i\eta) \\ -\lambda & 1-\frac{\lambda^2}{2} & A\lambda^2 \\ A\lambda^3(1-\rho-i\eta) & -A\lambda^2 & 1 \end{pmatrix} + \mathcal{O}(\lambda^4) \tag{1.30}$$

Equation 1.30 is truncated at $\mathcal{O}(\lambda^4)$, parameterisations at higher orders are [26]:

$$V_{\mathrm{CKM}} = \begin{pmatrix} 1-\frac{\lambda^2}{2}-\frac{\lambda^4}{8} & \lambda & A\lambda^3(\rho - i\eta) \\ -\lambda+\frac{1}{2}A^2\lambda^5[1-2(\rho+i\eta)] & 1-\frac{\lambda^2}{2}-\frac{1}{8}\lambda^4(1+4A^2) & A\lambda^2 \\ A\lambda^3[1-(1-\frac{1}{2}\lambda^2)(\rho+i\eta)] & -A\lambda^2+\frac{1}{2}A\lambda^4[1-2(\rho+i\eta)] & 1-\frac{1}{2}A^2\lambda^4 \end{pmatrix} + \mathcal{O}(\lambda^6) \tag{1.31}$$

(λ, A, ρ, η) are four real parameters, the latter three being of the order of one, and this representation is unitary to all orders of λ. Another useful feature of this parameterisation is that the quantity $(\rho - i\eta)$ is phase-convention independent and it is closely related to the amount of *CP* violation introduced in the SM, as described in Sect. 1.5.3. It should also be noted that at $\mathcal{O}(\lambda^3)$, the complex phase responsible for *CP* violation is present only in the far off-diagonal elements V_{ub} and V_{td}.

1.5.3 The Unitary Triangle (UT)

The unitarity of the CKM matrix V_{CKM} requires that:

$$V^*V = VV^* = \mathrm{I} \quad \Rightarrow \quad \sum_j V^*_{ji}V_{jk} = \sum_j V_{ij}V^*_{kj} = \delta_{ik} \tag{1.32}$$

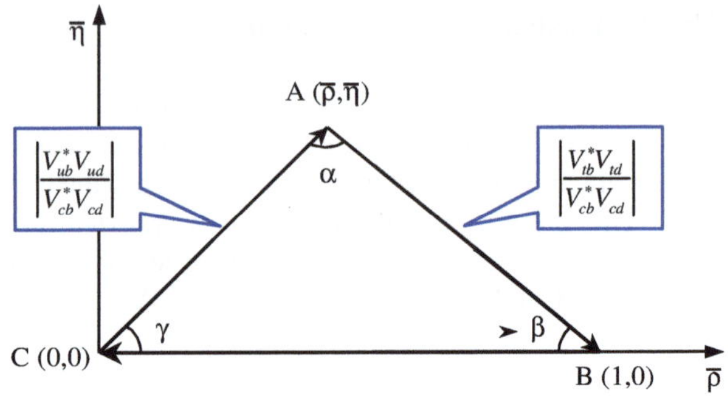

Fig. 1.3 The sketch of the unitarity triangle

These are nine independent equations: three of them are equal to one and six of them, deriving from the off-diagonal elements, are equal to zero. The six vanishing combinations can be represented as six triangles in a complex plane:

$$V_{cd}V_{ud}^* + V_{cs}V_{us}^* + V_{cb}V_{ub}^* = 0 \tag{1.33}$$
$$V_{cd}V_{td}^* + V_{cs}V_{ts}^* + V_{cb}V_{tb}^* = 0 \tag{1.34}$$
$$V_{ud}V_{td}^* + V_{us}V_{ts}^* + V_{ub}V_{tb}^* = 0 \tag{1.35}$$
$$V_{us}^*V_{ud} + V_{cs}^*V_{cd} + V_{ts}^*V_{td} = 0 \tag{1.36}$$
$$V_{ub}^*V_{us} + V_{cb}^*V_{cs} + V_{tb}^*V_{ts} = 0 \tag{1.37}$$
$$V_{ub}^*V_{ud} + V_{cb}^*V_{cd} + V_{tb}^*V_{td} = 0 \tag{1.38}$$

The areas of all these triangles are the same and equal to $J/2$, where J is the Jarlskog invariant [27] defined as $\mathrm{Im}[V_{ij}V_{kl}V_{il}^*V_{kj}^*] = J\sum_{m,n}\epsilon_{ikm}\epsilon_{jln}$, which gives a direct measure of the CP violation as introduced in the SM by the CKM mechanism. All the triangles are flat and nearly degenerate except those corresponding to Eq. 1.35 and Eq. 1.38. The latter is commonly used to represent the CP-violating phase content of Eq. 1.30. This is referred to as the Unitarity Triangle (UT). The UT is rotated and scaled such that $V_{cb}^*V_{cd}$ is real and $|V_{cb}^*V_{cd}| = 1$. This triangle, shown in Fig. 1.3, has two exact vertices at (0,0) and at (1,0) by construction and the coordinates $(\bar{\rho},\bar{\eta})$ of the remaining vertex are:

$$\bar{\rho} = \rho(1 - \lambda^2/2) \tag{1.39}$$
$$\bar{\eta} = \eta(1 - \lambda^2/2) \tag{1.40}$$

The three angles of the UT, denoted with α, β and γ, are defined as:

$$\alpha \equiv \arg\left[-\frac{V_{td}V_{tb}^*}{V_{ud}V_{ub}^*}\right] \quad \beta \equiv \arg\left[-\frac{V_{cd}V_{cb}^*}{V_{td}V_{tb}^*}\right] \quad \gamma \equiv \arg\left[-\frac{V_{ud}V_{ub}^*}{V_{cd}V_{cb}^*}\right] \tag{1.41}$$

A precise determination of the sides and the angles of the CKM matrix is crucial to test the CKM mechanism in the SM. Several independent measurements can be

Table 1.1 Fitted values to the Wolfenstein parameters and Jarlskog invariant

Parameter	Central $\pm 1\sigma$
A	$0.801^{+0.026}_{-0.014}$
λ	$0.22539^{+0.00062}_{-0.00095}$
ρ	$0.144^{+0.023}_{-0.026}$
η	$0.343^{+0.015}_{-0.014}$
$J(10^{-5})$	$2.884^{+0.253}_{-0.053}$

Table 1.2 Measured UT angles using direct measurements only

Parameter	Central $\pm 1\sigma$
α	$\left(89.0^{+4.4}_{-4.2}\right)^{\circ}$
β	$\left(21.38^{+0.79}_{-0.77}\right)^{\circ}$
γ	$\left(68^{+10}_{-11}\right)^{\circ}$

compared on the $(\bar{\rho}, \bar{\eta})$ plane and used to overconstrain the UT triangle: a detailed review on the measurements of each single CKM matrix element can be found elsewhere [2].

1.5.4 Global Fits to V_{CKM}

Each independent measurement of the parameters of V_{CKM} [2] can be used to test its unitary. The CKMfitter [28] group provides global fits to the UT including measurements involving both loop and tree diagrams. The best fits to the Wolfenstein parameters and Jarlskog invariant, taken from [28], are shown in Table 1.1, while the angles of the UT, using only direct measurements, are reported in Table 1.2.

At present, all direct measurements are consistent with expectations and global fits confirm a good agreement between data and the CKM mechanism (Fig. 1.4). The CKM angles α and β are measured with good accuracy, while the current value of γ is still affected by a large uncertainty. In particular, a good measurement of the angle γ is of particular interest since $\gamma \neq 0$ is a requirement in order to observe direct *CP* violation in tree decays, as shown in Sect. 1.6. In fact, γ is the only angle that does not involve any t quark coupling in its definition (Eq. 1.41) and provides the unique opportunity to be extracted from both tree and loop decays. A precise tree-level determination of γ could provide a clean benchmark to test the SM with respect to New Physics processes, which are expected to appear in loop-mediated decays.

It is also interesting to note that, in global fits, the UT is constrained mostly by loop-mediated processes (see Fig. 1.5). In fact, once tree level only processes are considered in the fit, the position of the apex of the triangle appears to be much less defined in the $(\bar{\rho},\bar{\eta})$ plane.

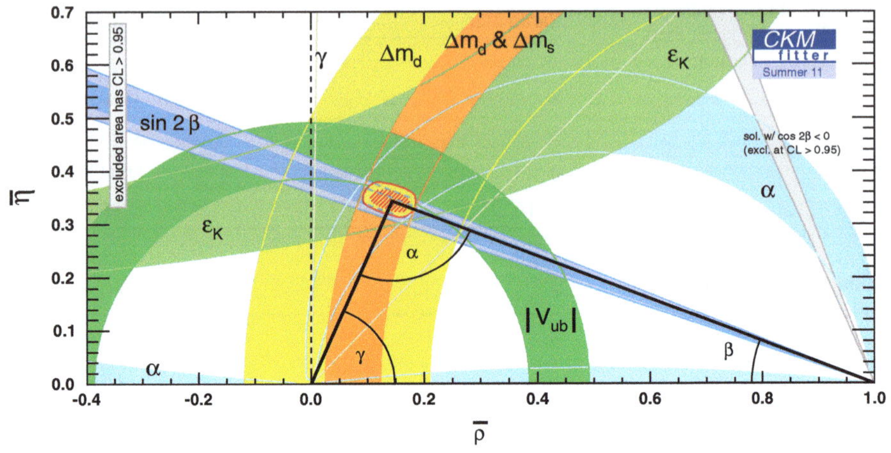

Fig. 1.4 Constraints on the $(\bar{\rho},\bar{\eta})$ plane. The vertex of the UT appears well constrained

1.6 Tree Level Determination of γ with $B \to DK$ Decays

The CKM angle γ, defined in Eq. 1.41, is essentially the phase of V_{ub}. It can be extracted in $B \to DK$ decays at tree-level exploiting the interference between $b \to c(\bar{u}s)$ and $b \to u(\bar{c}s)$ transitions. This was first proposed by M. Gronau and D. London [29] and M. Gronau and D. Wyler [30]. Charged $B^{\pm} \to DK^{\pm}$ decays can proceed through two different amplitudes, shown in Fig. 1.6.

$B^- \to D^0 K^-$ and $B^- \to \bar{D}^0 K^-$ amplitudes (charge conjugation implied) can interfere when both D^0 and $\bar{D}^0$ flavour eigenstates decay into the same final state f. Noting that the phase information in Eq. 1.30 is contained within the V_{ub} element and the V_{cb} element has no complex phase, $V_{ub} \propto e^{-i\gamma}$ and the amplitudes involved in the decay can be written as:

$$A(B^- \to D^0 K^-) \propto 1 \tag{1.42}$$
$$A(B^- \to \bar{D}^0 K^-) \propto r_B e^{i(\delta_B - \gamma)} \tag{1.43}$$
$$A(B^+ \to \bar{D}^0 K^+) \propto 1 \tag{1.44}$$
$$A(B^+ \to D^0 K^+) \propto r_B e^{i(\delta_B + \gamma)} \tag{1.45}$$

where all equations have been normalized to the corresponding favoured amplitudes (the favoured amplitudes are proportional to 1 by construction). This set of equations depends on three real parameters: the modulus of the ratio of the suppressed/favoured amplitudes $r_B = |A_{\text{sup}}/A_{\text{fav}}|$, a *CP*-invariant *strong* phase δ_B and the *CP*-violating weak phase γ. Note that only γ changes its relative sign under *CP* transformations. Assuming negligible *CP* violation and mixing in the D^0 system and with the following definitions:

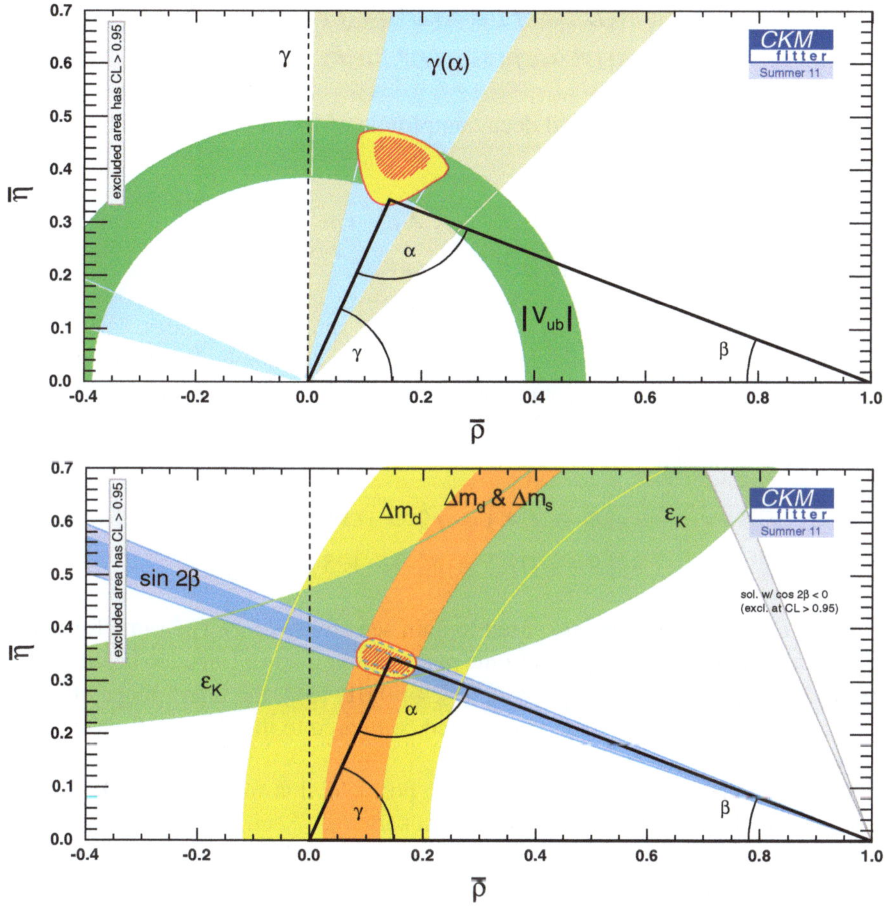

Fig. 1.5 Global fit to the UT. *Left* constraints from "tree" quantities only. *Right* constraints from "loop" quantities only

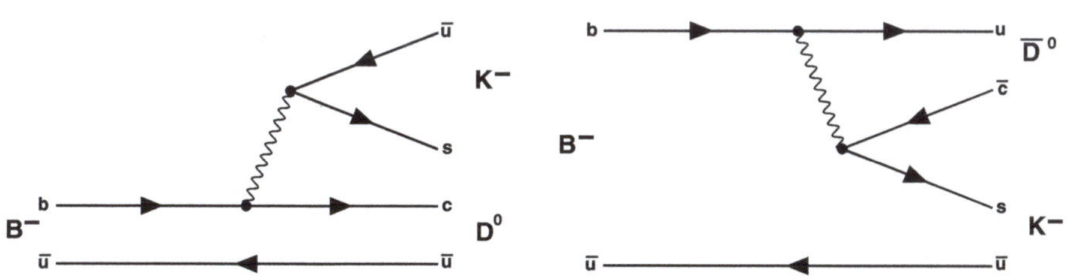

Fig. 1.6 Feynman diagrams for $B^{\pm} \to DK^{\pm}$ decays. Both processes proceed via tree-level diagrams. *Left* colour favoured amplitude A_{fav}. *Right* colour suppressed amplitude A_{sup}. r_b defines the ratio of the two amplitudes and it is defined as $r_B = |A_{\text{sup}}/A_{\text{fav}}|$

$$A(D^0 \to f) \equiv A(\bar{D}^0 \to \bar{f}) = A_f \tag{1.46}$$
$$A(\bar{D}^0 \to f) \equiv A(D^0 \to \bar{f}) = \bar{A}_f \tag{1.47}$$

it is possible to express the total decay amplitude of the $B \to [f]_D K$ chain ($B \to DK$ with $D \to f$) as:

$$A(B^- \to [f]_D K^-) \propto A_f + r_B e^{i(\delta_B - \gamma)} \bar{A}_f \tag{1.48}$$
$$A(B^+ \to [\bar{f}]_D K^+) \propto r_B e^{i(\delta_B + \gamma)} A_f + \bar{A}_f \tag{1.49}$$

where D denotes the superposition of D^0 and $\bar{D}^0$ intermediate states. The interference between those two amplitudes can give rise to detectable direct *CP* violation, although r_B is small due to colour suppression and therefore interference effects are expected to be in general small. For $B \to DK$ decays the best measured values of r_B and δ_B are [2, 31]:

$$r_B[B \to DK] \equiv r_B^K = 0.113^{+0.024}_{-0.021} \tag{1.50}$$
$$\delta_B[B \to DK] \equiv \delta_B^K = (125 \pm 16)^\circ \tag{1.51}$$

$B \to D\pi$ decays can be also used to gain sensitivity to γ. The same formalism applies, although two different hadronic parameters r_B^π and δ_B^π need to be introduced for the $B \to D\pi$ mode. As $r_B^\pi \ll r_B^K$ ($r_B^\pi \approx 0.01$) [31], the interference effects in $B \to D\pi$ decays are expected to be limited compared to the $B \to DK$ case, and so is the sensitivity to γ.

Several choices of the D final state f are possible: this analysis uses $D \to h^+h^-$ two-body final states, where h is a pion or a kaon (see Sect. 3.1) Two-body charged final states provide higher reconstruction efficiencies in the LHCb detector and lower combinatorial background compared to multi-body modes, thus being the easier choice for this type of analyses.

1.6.1 GLW Method

he GLW method takes its name from the initials of its proponents M. Gronau, D. London and D. Wyler [29, 30]. The original idea considers $D \to f$ final states where f is an exact *CP* eigenstate ($f = f_{CP}$). In this hypothesis, Eq. 1.48 and Eq. 1.49 factorise into:

$$A(B^- \to [f_{CP}]_D K^-) \propto 1 + r_B e^{i(\delta_B - \gamma)} \tag{1.52}$$
$$A(B^+ \to [f_{CP}]_D K^+) \propto r_B e^{i(\delta_B + \gamma)} + 1 \tag{1.53}$$

and can be used to calculate the partial widths:

$$\Gamma(B^- \to [f_{CP}]_D K^-) \propto |A(B^- \to DK^-)|^2 \propto 1 + r_B^2 + 2r_B \cos(\delta_B - \gamma) \tag{1.54}$$

$$\Gamma(B^+ \to [f_{CP}]_D K^+) \propto |A(B^+ \to DK^+)|^2 \propto 1 + r_B^2 + 2r_B \cos(\delta_B + \gamma) \tag{1.55}$$

The two amplitudes involved differ in size with the ratio r_B, so interference is limited. As in the LHCb hadronic environment neutral final states are difficult to reconstruct (e.g. the CP-odd $D \to K_s\pi^0$ final state), the most accessible modes are the two-body CP-even final states, $D \to K^+K^-$ and $D \to \pi^+\pi^-$.

1.6.2 ADS Method

Proposed by D. Atwood, I. Dunietz and A. Soni [32, 33], the ADS method suggests the use of non-CP eigenstates like $D \to K^\pm\pi^\mp$. In this case, $D \to K^\pm\pi^\mp$ decays proceed via two different amplitudes as illustrated in Fig. 1.7: either a Cabibbo-favoured decay $\bar{D}^0 \to K^+\pi^-$ or a doubly-Cabibbo-suppressed (DCS) decay $D^0 \to K^+\pi^-$ (charge conjugation implied).

Analogously, these amplitudes can be parameterised as[4]:

$$A(\bar{D}^0 \to K^+\pi^-) \propto 1 \tag{1.56}$$

$$A(D^0 \to K^+\pi^-) \propto r_D e^{-i\delta_D} \tag{1.57}$$

with the introduction of the D-amplitude ratio r_D and the strong phase δ_D (no CP-violating phase assumed here). Both r_D and δ_D depend on the particular $D \to f$ final state and in this case they will indicate $f = K^\pm\pi^\mp$, unless otherwise stated. The reverse suppression sketched in Fig. 1.8 can be used to enhance the interference effects, as the two resulting amplitudes are comparable in size. However, the total branching fraction of the suppressed $B \to DK$ decays is expected to be small, as

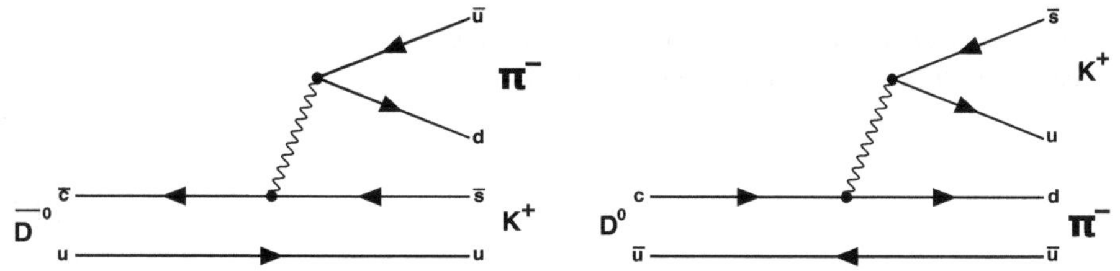

Fig. 1.7 Feynman diagrams for $D \to K\pi$ decays. *Left* Cabibbo-favoured amplitude. *Right* doubly-Cabibbo-suppressed amplitude

[4] Note that the strong phase in the suppressed D decay carries a minus sign, unlike the definition in the B decay.

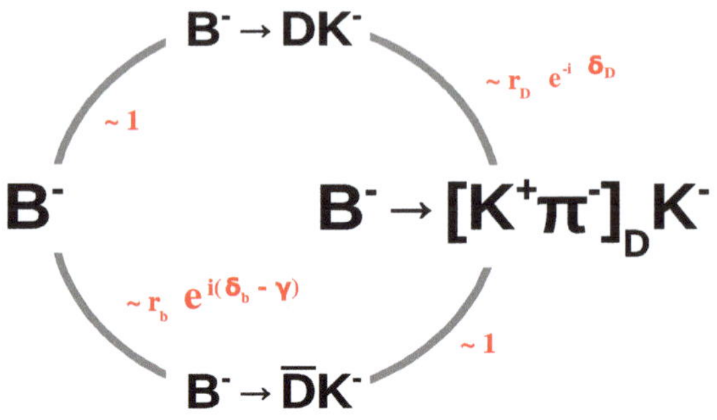

Fig. 1.8 The ADS method. Favoured $B^- \to D^0K^-$ decays are followed by DCS $D^0 \to K^+\pi^-$ decays. Vice versa, colour suppressed $B^- \to \bar{D}^0K^-$ decays are followed by Cabibbo-favoured $\bar{D}^0 \to K^+\pi^-$ transitions

low as 10^{-7} [34]. The formulas for the ADS partial decay widths are similar to those presented in Sect. 1.6.1, with the only addition of the extra dependence on the two parameters r_D and δ_D. The total decay amplitudes, including both favoured and suppressed modes, are:

$$A(B^- \to [K^+\pi^-]_D K^-) \propto r_D e^{-i\delta_D} + r_B e^{i(\delta_B - \gamma)} \tag{1.58}$$

$$A(B^+ \to [K^-\pi^+]_D K^+) \propto r_B e^{i(\delta_B + \gamma)} + r_D e^{-i\delta_D} \tag{1.59}$$

$$A(B^- \to [K^-\pi^+]_D K^-) \propto 1 + r_B r_D e^{i(\delta_B - \gamma)} e^{-i\delta_D} \tag{1.60}$$

$$A(B^+ \to [K^+\pi^-]_D K^+) \propto 1 + r_B r_D e^{i(\delta_B + \gamma)} e^{-i\delta_D} \tag{1.61}$$

All the partial widths of the suppressed decay can be expressed as:

$$\Gamma(B^- \to [K^+\pi^-]_D K^-) \propto r_D^2 + r_B^2 + 2r_D r_B \cos(\delta_B + \delta_D - \gamma) \tag{1.62}$$

$$\Gamma(B^+ \to [K^-\pi^+]_D K^+) \propto r_D^2 + r_B^2 + 2r_D r_B \cos(\delta_B + \delta_D + \gamma) \tag{1.63}$$

as well as those of the favoured mode:

$$\Gamma(B^- \to [K^-\pi^+]_D K^-) \propto 1 + r_D^2 r_B^2 + 2r_D r_B \cos(\delta_B - \delta_D - \gamma) \tag{1.64}$$

$$\Gamma(B^+ \to [K^+\pi^-]_D K^+) \propto 1 + r_D^2 r_B^2 + 2r_D r_B \cos(\delta_B - \delta_D + \gamma) \tag{1.65}$$

The r_D and δ_D parameters can be measured independently in tagged $D \to K\pi$ decays. Their best measured values are [31]:

$$r_D = (6.16 \pm 0.15)\% \tag{1.66}$$

$$\delta_D = \left(202.4^{+9.7}_{-11.0}\right)^\circ \tag{1.67}$$

1.6.3 Physical Observables

The formulas introduced in Sect. 1.6.1 and Sect. 1.6.2 can be used to construct useful, *CP*-related, physical observables, which depend on the r, δ and γ parameters introduced before. The most accessible experimental quantities that can be measured are the ratios of branching fractions $R_{K/\pi}$ and the charge asymmetries A, defined for each mode as:

$$R_{K/\pi} = \frac{\Gamma(B^- \to DK^-) + \Gamma(B^+ \to DK^+)}{\Gamma(B^- \to D\pi^-) + \Gamma(B^+ \to D\pi^+)} \tag{1.68}$$

$$A = \frac{\Gamma(B^- \to Dh^-) - \Gamma(B^+ \to Dh^+)}{\Gamma(B^- \to Dh^-) + \Gamma(B^+ \to Dh^+)} \tag{1.69}$$

Those quantities are particularly useful as they can be measured with simple *counting* experiments. Furthermore, most of the multiplicative systematic uncertainties cancel in ratios by construction and only small corrections are required (see Sect. 5.3.1) For the GLW modes $B^\pm \to [K^+K^-, \pi^+\pi^-]_D h^\pm$, the following quantities are of interest:

$$A_{CP+} = \frac{\Gamma(B^- \to D_{CP+}K^-) - \Gamma(B^+ \to D_{CP+}K^+)}{\Gamma(B^- \to D_{CP+}K^-) + \Gamma(B^+ \to D_{CP+}K^+)} \tag{1.70}$$

$$R_{CP+} = 2\frac{\Gamma(B^- \to D_{CP+}K^-) + \Gamma(B^+ \to D_{CP+}K^+)}{\Gamma(B^- \to D^0K^-) + \Gamma(B^+ \to \bar{D}^0K^+)} \tag{1.71}$$

where D_{CP+} indicates the $D \to K^+K^-$ and $D \to \pi^+\pi^-$ final states. On an experimental point of view, it is convenient to measure the following quantities:

$$\frac{\Gamma(B^- \to [K^-\pi^+]_D K^-) + \Gamma(B^+ \to [K^+\pi^-]_D K^+)}{\Gamma(B^- \to [K^-\pi^+]_D \pi^-) + \Gamma(B^+ \to [K^+\pi^-]_D \pi^+)} = R^{K\pi}_{K/\pi} \tag{1.72}$$

$$\frac{\Gamma(B^- \to [K^-K^+]_D K^-) + \Gamma(B^+ \to [K^+K^-]_D K^+)}{\Gamma(B^- \to [K^-K^+]_D \pi^-) + \Gamma(B^+ \to [K^+K^-]_D \pi^+)} = R^{KK}_{K/\pi} \tag{1.73}$$

$$\frac{\Gamma(B^- \to [\pi^-\pi^+]_D K^-) + \Gamma(B^+ \to [\pi^+\pi^-]_D K^+)}{\Gamma(B^- \to [\pi^-\pi^+]_D \pi^-) + \Gamma(B^+ \to [\pi^+\pi^-]_D \pi^+)} = R^{\pi\pi}_{K/\pi} \tag{1.74}$$

$$\frac{\Gamma(B^- \to [K^-\pi^+]_D \pi^-) - \Gamma(B^+ \to [K^+\pi^-]_D \pi^+)}{\Gamma(B^- \to [K^-\pi^+]_D \pi^-) + \Gamma(B^+ \to [K^+\pi^-]_D \pi^+)} = A^{K\pi}_{\pi} \tag{1.75}$$

$$\frac{\Gamma(B^- \to [K^-\pi^+]_D K^-) - \Gamma(B^+ \to [K^+\pi^-]_D K^+)}{\Gamma(B^- \to [K^-\pi^+]_D K^-) + \Gamma(B^+ \to [K^+\pi^-]_D K^+)} = A^{K\pi}_{K} \tag{1.76}$$

$$\frac{\Gamma(B^- \to [K^-K^+]_D K^-) - \Gamma(B^+ \to [K^+K^-]_D K^+)}{\Gamma(B^- \to [K^-K^+]_D K^-) + \Gamma(B^+ \to [K^+K^-]_D K^+)} = A^{KK}_{K} \tag{1.77}$$

$$\frac{\Gamma(B^- \to [\pi^-\pi^+]_D K^-) - \Gamma(B^+ \to [\pi^+\pi^-]_D K^+)}{\Gamma(B^- \to [\pi^-\pi^+]_D K^-) + \Gamma(B^+ \to [\pi^+\pi^-]_D K^+)} = A^{\pi\pi}_{K} \tag{1.78}$$

$$\frac{\Gamma(B^- \to [K^-K^+]_D \pi^-) - \Gamma(B^+ \to [K^+K^-]_D \pi^+)}{\Gamma(B^- \to [K^-K^+]_D \pi^-) + \Gamma(B^+ \to [K^+K^-]_D \pi^+)} = A^{KK}_{\pi} \tag{1.79}$$

$$\frac{\Gamma(B^- \to [\pi^-\pi^+]_D \pi^-) - \Gamma(B^+ \to [\pi^+\pi^-]_D \pi^+)}{\Gamma(B^- \to [\pi^-\pi^+]_D \pi^-) + \Gamma(B^+ \to [\pi^+\pi^-]_D \pi^+)} = A^{\pi\pi}_{\pi} \tag{1.80}$$

These experimental observables can be related to the r_B, δ_B and γ parameters using the equations in Sect. 1.6.1 and Sect. 1.6.2:

$$R_{K/\pi}^{K\pi} = R\frac{1 + (r_B^K r_D)^2 + 2r_B^K r_D \cos(\delta_B^K - \delta_D)\cos\gamma}{1 + (r_B^\pi r_D)^2 + 2r_B^\pi r_D \cos(\delta_B^\pi - \delta_D)\cos\gamma} \tag{1.81}$$

$$R_{K/\pi}^{KK} = R_{K/\pi}^{\pi\pi} = R\frac{1 + {r_B^K}^2 + 2r_B^K \cos\delta_B^K \cos\gamma}{1 + {r_B^\pi}^2 + 2r_B^\pi \cos\delta_B^\pi \cos\gamma} \tag{1.82}$$

$$A_K^{K\pi} = \frac{2r_B^K r_D \sin(\delta_B^K - \delta_D)\sin\gamma}{1 + (r_B^K r_D)^2 + 2r_B^K r_D \cos(\delta_B^K - \delta_D)\cos\gamma} \tag{1.83}$$

$$A_\pi^{K\pi} = \frac{2r_B^\pi r_D \sin(\delta_B^\pi - \delta_D)\sin\gamma}{1 + (r_B^\pi r_D)^2 + 2r_B^\pi r_D \cos(\delta_B^\pi - \delta_D)\cos\gamma} \tag{1.84}$$

$$A_K^{KK} = A_K^{\pi\pi} = \frac{2r_B^K \sin\delta_B^K \sin\gamma}{1 + {r_B^K}^2 + 2r_B^K \cos\delta_B^K \cos\gamma} \tag{1.85}$$

$$A_\pi^{KK} = A_\pi^{\pi\pi} = \frac{2r_B^\pi \sin\delta_B^\pi \sin\gamma}{1 + {r_B^\pi}^2 + 2r_B^\pi \cos\delta_B^\pi \cos\gamma} \tag{1.86}$$

where the superscript K or π on the r_B and δ_B parameters indicates $B \to DK$ and $B \to D\pi$ modes respectively. In detail:

$$R = \frac{\Gamma(B^- \to D^0 K^-)}{\Gamma(B^- \to D^0 \pi^-)} \tag{1.87}$$

$$r_B^K = \left|\frac{A(B^- \to \overline{D}^0 K^-)}{A(B^- \to D^0 K^-)}\right| \tag{1.88}$$

$$r_B^\pi = \left|\frac{A(B^- \to \overline{D}^0 \pi^+)}{A(B^- \to D^0 \pi^+)}\right| \tag{1.89}$$

$$r_D = \left|\frac{A(\overline{D}^0 \to K^- \pi^+)}{A(D^0 \to K^- \pi^+)}\right| \tag{1.90}$$

$$\delta_B^K = \arg(B^- \to \overline{D}^0 K^+) - \arg(B^- \to D^0 K^+) \tag{1.91}$$

$$\delta_B^\pi = \arg(B^- \to \overline{D}^0 \overset{+}{\pi}) - \arg(B^- \to D^0 \overset{+}{\pi}) \tag{1.92}$$

$$\delta_D = \arg(D^0 \to K^- \overset{+}{\pi}) - \arg(\overline{D}^0 \to K^- \overset{+}{\pi}) \tag{1.93}$$

It is experimentally convenient to approximate R_{CP+} as the double ratio:

$$R_{CP+} \approx \frac{R_{K/\pi}^{KK,\pi\pi}}{R_{K/\pi}^{K\pi}} \tag{1.94}$$

This approximation assumes no *CP* violation in both $B \to [KK, \pi\pi]_D\pi$ modes nor in the favoured $B \to DK$ mode ($r_B^\pi \ll 1$ and $r_B^K r_D \ll 1$). It is particularly convenient as no systematic uncertainty coming from differences in the reconstruction efficiencies of the various $D \to hh$ modes need to be considered (the dependence on the D final state is removed). Under this approximation Eq. 1.70 and Eq. 1.71 become:

$$A_{CP+} = \frac{2r_B^K \sin\delta_B^K \sin\gamma}{1 + {r_B^K}^2 + 2r_B^K \cos\delta_B^K \cos\gamma} = A_K^{KK} = A_K^{\pi\pi} \tag{1.95}$$

$$R_{CP+} = 1 + {r_B^K}^2 + 2r_B^K \cos\delta_B^K \cos\gamma \approx \frac{R_{K/\pi}^{KK,\pi\pi}}{R_{K/\pi}^{K\pi}} \tag{1.96}$$

with the last columns representing the quantities experimentally measured.
For the suppressed $B^\pm \to [K^\mp\pi^\pm]_D h^\pm$ decays, the following *ADS* quantities can be introduced:

$$\frac{\Gamma(B^- \to [K^+\pi^-]_D K^-) + \Gamma(B^+ \to [K^-\pi^+]_D K^+)}{\Gamma(B^- \to [K^-\pi^+]_D K^-) + \Gamma(B^+ \to [K^+\pi^-]_D K^+)} = R_{\text{ADS}(K)} \tag{1.97}$$

$$\frac{\Gamma(B^- \to [K^+\pi^-]_D \pi^-) + \Gamma(B^+ \to [K^-\pi^+]_D \pi^+)}{\Gamma(B^- \to [K^-\pi^+]_D \pi^-) + \Gamma(B^+ \to [K^+\pi^-]_D \pi^+)} = R_{\text{ADS}(\pi)} \tag{1.98}$$

$$\frac{\Gamma(B^- \to [K^+\pi^-]_D K^-) - \Gamma(B^+ \to [K^-\pi^+]_D K^+)}{\Gamma(B^- \to [K^+\pi^-]_D K^-) + \Gamma(B^+ \to [K^-\pi^+]_D K^+)} = A_{\text{ADS}(K)} \tag{1.99}$$

$$\frac{\Gamma(B^- \to [K^+\pi^-]_D \pi^-) - \Gamma(B^+ \to [K^-\pi^+]_D \pi^+)}{\Gamma(B^- \to [K^+\pi^-]_D \pi^-) + \Gamma(B^+ \to [K^-\pi^+]_D \pi^+)} = A_{\text{ADS}(\pi)} \tag{1.100}$$

$A_{\text{ADS}(\pi)}$ and $A_{\text{ADS}(K)}$ are the charge asymmetries measured in the suppressed $K\pi$ modes, while $R_{\text{ADS}(K)}$ and $R_{\text{ADS}(\pi)}$ are the ratios of the suppressed/favoured branching fractions. Analogously, these quantities can be rewritten as:

$$R_{\text{ADS}(K)} = \frac{{r_B^K}^2 + {r_D}^2 + 2r_B^K r_D \cos(\delta_B^K + \delta_D)\cos\gamma}{1 + (r_B^K r_D)^2 + 2r_B^K r_D \cos(\delta_B^K - \delta_D)\cos\gamma} \tag{1.101}$$

$$A_{\text{ADS}(K)} = \frac{2r_B^K r_D \sin(\delta_B^K + \delta_D)\sin\gamma}{{r_B^K}^2 + {r_D}^2 + 2r_B^K r_D \cos(\delta_B^K + \delta_D)\cos\gamma} \tag{1.102}$$

$$R_{\text{ADS}(\pi)} = \frac{{r_B^\pi}^2 + {r_D}^2 + 2r_B^\pi r_D \cos(\delta_B^\pi + \delta_D)\cos\gamma}{1 + (r_B^\pi r_D)^2 + 2r_B^\pi r_D \cos(\delta_B^\pi - \delta_D)\cos\gamma} \tag{1.103}$$

$$R_{\text{ADS}(\pi)} = \frac{2r_B^\pi r_D \sin(\delta_B^\pi + \delta_D)\sin\gamma}{{r_B^\pi}^2 + {r_D}^2 + 2r_B^\pi r_D \cos(\delta_B^\pi + \delta_D)\cos\gamma} \tag{1.104}$$

The pair of variables $[R_{\text{ADS}(h)}, A_{\text{ADS}(h)}]$ can be replaced with a second pair of variables $[R_h^+, R_h^-]$ defined as:

$$R_K^- = \frac{\Gamma(B^- \to [\pi^- K^+]_D K^-)}{\Gamma(B^- \to [K^- \pi^+]_D K^-)} \tag{1.105}$$

$$R_K^+ = \frac{\Gamma(B^+ \to [\pi^+ K^-]_D K^+)}{\Gamma(B^+ \to [K^+ \pi^-]_D K^+)} \tag{1.106}$$

$$R_\pi^- = \frac{\Gamma(B^- \to [\pi^- K^+]_D \pi^-)}{\Gamma(B^- \to [K^- \pi^+]_D \pi^-)} \tag{1.107}$$

$$R_\pi^+ = \frac{\Gamma(B^+ \to [\pi^+ K^-]_D \pi^+)}{\Gamma(B^+ \to [K^+ \pi^-]_D \pi^+)} \tag{1.108}$$

It is possible to move from one set of variables to another with simple linear transformations:

$$R_{\text{ADS}(h)} = \frac{R_h^- + R_h^+}{2} \tag{1.109}$$

$$A_{\text{ADS}(h)} = \frac{R_h^- - R_h^+}{R_h^- + R_h^+} \tag{1.110}$$

Unlike $R_{\text{ADS}(h)}$ and $A_{\text{ADS}(h)}$, R_h^+ and R_h^- are by construction statistically uncorrelated, making the combination of errors easier when the results are used in the fit to γ (see Sect. 6.6).

These observables have been studied before both at the b-factories and the Tevatron. A detailed comparison of the latest results can be found in Sect. 6.5.

References

1. F. Halzen, A.D. Martin, *Quarks and Leptons: An Introductory Course in Modern Particle Physics*. (Wiley & Sons, New York, 1984)
2. Particle Data Group, Nakamura, K et al., Review of particle physics, J. Phys. **G37**, 075021 (2010)
3. M.E. Peskin, D.V. Schroeder, *An Introduction to Quantum Field Theory*. (Harper-Collins, New York, 1995)
4. W. Greiner, S. Schramm, E. Stein, *Quantum Chromodynamics*. (Springer, Berlin, 2007)
5. F. Englert, R. Brout, Broken symmetry and the mass of gauge vector mesons. Phys. Rev. Lett. **13**, 321 (1964)
6. P. W. Higgs, Broken symmetries and the masses of gauge bosons. Phys. Rev. Lett. **13**, 508 (1964)
7. A.D. Sakharov, Violation of CP invariance, C asymmetry, and baryon asymmetry of the universe. J. Exp. Theor. Phys. **5**, 24–27 (1967)
8. E. Noether, Invariante variations probleme (Math-phys, Nachr. D. König. Gesellsch. D. Wiss. Zu Göttingen, 1918), p. 235
9. H.P. Strapp, Derivation of the CPT theorem and the connection between spin and statistics from postulates of the S-Matrix theory. Phys. Rev. Lett. **125**, 2139 (1962)
10. O.W. Greenberg, CPT Violation implies violation of Lorentz invariance. Phys. Rev. Lett. **89**, 231602 (2002)

11. C.S. Wu et al., Experimental test of parity conservation in beta decay. Phys. Rev. **105**, 1413 (1957)
12. J.H. Christenson, J.W. Cronin, V.L. Fitch, R. Turlay, Evidence for the 2π decay of the k(2)0 meson. Phys. Rev. Lett. **13**, 138 (1964)
13. The BABAR collaboration, Aubert, B et al., The BaBar detector. Nucl. Instrum. Meth. **A479**, 1 (2002) arXiv:hep-ex/0105044
14. A. Abashian et al., The belle detector. Nucl. Instrum. Meth. **A479**, 117 (2002)
15. R. Blair et al., The CDF-II detector: technical design report. Fermilab PUB-96-390-E (1996)
16. A. Abachi et al., The D0 Upgrade, Fermilab PUB-96-357-E (1996)
17. A. Angelopoulos et al., First direct observation of time-reversal non-invariance in the neutral-kaon system. Phys. Lett. B **444**(12), 43 (1998)
18. M. Gell-Mann, A. Pais, Behavior of neutral particles under charge conjugation. Phys. Rev. **97**, 1387 (1955)
19. M. Kobayashi, T. Maskawa, CP violation in the renormalizable theory of weak interaction. Prog. Theor. Phys. **49**, 652 (1973)
20. N. Cabibbo, Unitary symmetry and leptonic decays. Phys. Rev. Lett. **10**, 531 (1963)
21. S.L. Glashow, J. Iliopoulos, L. Maiani, Weak interactions with Lepton-Hadron symmetry. Phys. Rev. D **2**, 1285 (1970)
22. J. Aubert et al., Experimental observation of a heavy particle J. Phys. Rev. Lett. **33**, 1404 (1974)
23. S.W. Herb et al., Observation of a dimuon resonance at 9.5 GeV in 400-GeV proton-nucleus collisions. Phys. Rev. Lett. **39**, 252 (1977)
24. L.-L. Chau, W.-Y. Keung, Comments on the parametrization of the Kobayashi-Maskawa matrix. Phys. Rev. Lett. **53**, 1802 (1984)
25. Lincoln Wolfenstein, Parametrization of the Kobayashi-Maskawa matrix. Phys. Rev. Lett. **51**, 1945 (1983)
26. A.J. Buras, M.E. Lautenbacher, G. Ostermaier, Waiting for the top quark mass, $K^+ \rightarrow \pi^+ \nu \bar{nu}$, $B_s^0 - \bar{B}_s^0$ mixing, and CP asymmetries in B decays. Phys. Rev. D **50**, 3433 (1994)
27. C. Jarlskog, Commutator of the quark mass matrices in the standard electroweak model and a measure of maximal CP nonconservation. Phys. Rev. Lett. **55**, 1039 (1985)
28. CKM fitter group, J. Charles et al., Predictions of selected flavor observables within the standard model. Phys. Rev. D **84**, 033005 (2011)
29. M. Gronau, D. London, How to determine all the angles of the unitarity triangle from $B_d^0 \rightarrow DKS$ and $B_s^0 \rightarrow Dphi$. Phys. Lett. **B253**, 483 (1991)
30. M. Gronau, D. Wyler, On determining a weak phase from CP asymmetries in charged B decays. Phys. Lett. **B265**, 172 (1991)
31. Heavy Flavor Averaging Group, Y. Amhis et al., Averages of b-hadron, c-hadron, and tau-lepton properties as of early 2012, arXiv:1207.1158
32. D. Atwood, I. Dunietz, A. Soni, Enhanced CP violation with $B \rightarrow KD^0(\bar{D}^0)$ modes and extraction of the CKM angleγ. Phys. Rev. Lett. **78**, 3257 (1997) http://xxx.lanl.gov/abs/hep-ph/9612433
33. D. Atwood, I. Dunietz, A. Soni, Improved methods for observing CP violation in $BpmtoKD$ and measuring the CKM phase g. Phys. Rev. **D63**, 036005 (2001) http://xxx.lanl.gov/abs/hep-ph/0008090
34. Belle collaboration, Y. Horii et al., Evidence for the suppressed decay$B \rightarrow DK^-, D \rightarrow K^+\pi^-$. Phys. Rev. Lett. **106**, 231803 (2011) http://arXiv:1103.5951

Chapter 2
The LHCb Experiment at the LHC

The purpose of this chapter is to give a brief introduction to the LHCb detector, the experimental apparatus used to perform the analysis described in this thesis. The chapter starts with a description of the Large Hadron Collider (LHC) at CERN in Sect. 2.1, while the detector is described in Sect. 2.2. After an overview of the running conditions of the experiment in 2011 (Sect. 2.2.1), all subdetectors are described individually in Sect. 2.2.2 to Sect. 2.2.7. The trigger system is described in Sect. 2.2.8, focusing the attention on the hadronic lines used in the analysis (for a detailed description of trigger requirements, see Sect. 3.6). The chapter is then concluded by a description of the track reconstruction in Sect. 2.2.9, the definition of the combined particle identification (PID) likelihood variables in Sect. 2.2.10 and an overview of the LHCb software framework used in the analysis in Sect. 2.2.11.

2.1 The LHC at CERN

The LHCb detector is one of the main experiments at the Large Hadron Collider (LHC) [1]. The LHC is a *proton-proton* collider located at CERN (Conseil Européen pour la Recherche Nucléaire), situated on the Franco-Swiss border near the city of Geneva. With its design energy of $\sqrt{s}$ =14 TeV and a design luminosity of $10^{34}\mathrm{cm}^{-2}\mathrm{s}^{-1}$, it is the largest and the most powerful particle collider ever built. Approved in December 1994 and constructed from 1998 to 2008, this superconductive synchrotron is situated in the same 26.7 km tunnel used for the e^+e^- collider LEP, decommissioned in 2000 after more than 10 years of successful running. A schematic view of the LHC is presented in Figs. 2.1 and 2.2. The accelerator is divided in eight octants and beams circulate in opposite directions into two separate rings. Two separate rings, with opposite magnetic field, are required as the colliding particles carry the same charge. This has direct consequences on the design of the magnets, shown in Fig. 2.3, which excludes a common vacuum beampipe for both circulating rings. All superconductive magnets are cooled down to a

P. Gandini, *Observation of CP Violation in $B^\pm \to DK^\pm$ Decays*,
Springer Theses, DOI: 10.1007/978-3-319-01029-8_2,

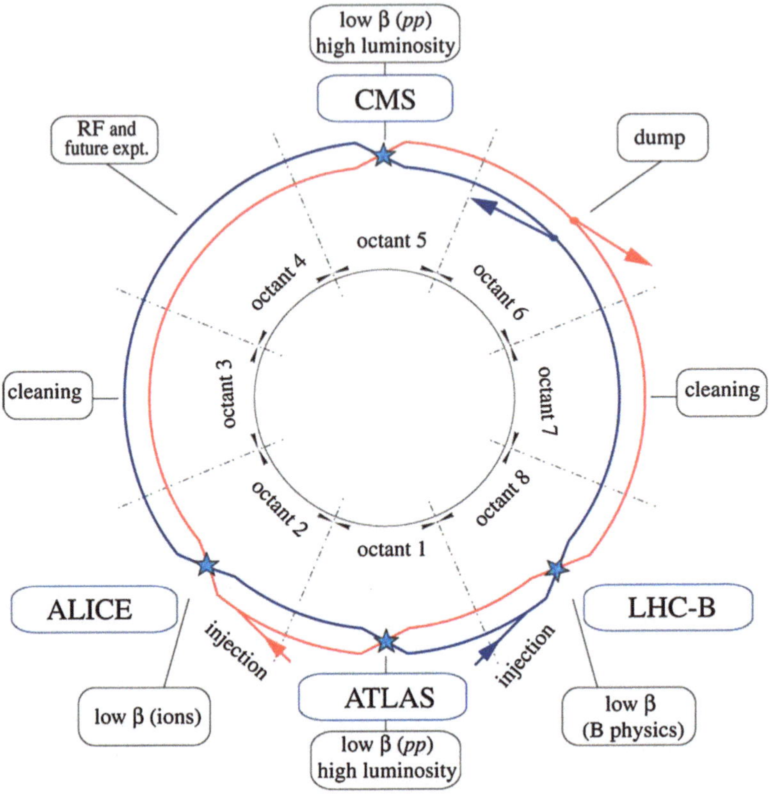

Fig. 2.1 A schematic layout of the LHC. Beam 1 runs clockwise, beam 2 anticlockwise

temperature of 1.9 K, using superfluid helium, to operate them safely at magnetic fields up to 8.3 T, depending on the energy of the beams. Protons are injected from the Super Proton Synchrotron (SPS) at the energy of 450 GeV and then accelerated to 3.5 TeV, providing collision at the total energy $\sqrt{s} = 7$ TeV. This corresponds to half of the design energy, which will be reached after upgrading the machine in 2013. Beams are kept separate in the machine and they only cross in four interaction points (IP), where the main experimental detectors are situated: ATLAS [2] and CMS [3] are hosted in IP1 and IP5 respectively, LHCb [4] in IP8 and ALICE [5] in IP2. The general purpose detectors ATLAS and CMS are designed to run at the highest possible peak luminosity of $10^{34}\,\text{cm}^{-2}\text{s}^{-1}$, while LHCb is designed to operate at a reduced instantaneous luminosity of $10^{32}\,\text{cm}^{-2}\text{s}^{-1}$ (for a detailed description of the running conditions of the LHCb detector, see Sect. 2.2.1). The high luminosity of the LHC is delivered through intense bunches consisting of 1.1×10^{11} protons each, separated by 50 ns intervals between each crossing (the design value is 25 ns corresponding to a frequency of 40 MHz).

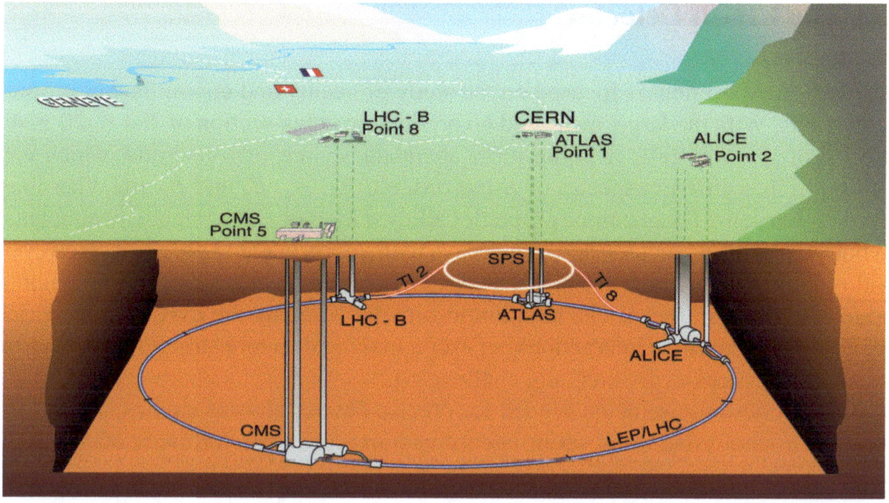

Fig. 2.2 A pictorial view of the LHC collider with its four interaction points The underground tunnel lies between 50 and 175 m below the surface on a plane inclined down about 1.4 % towards Lac Léman

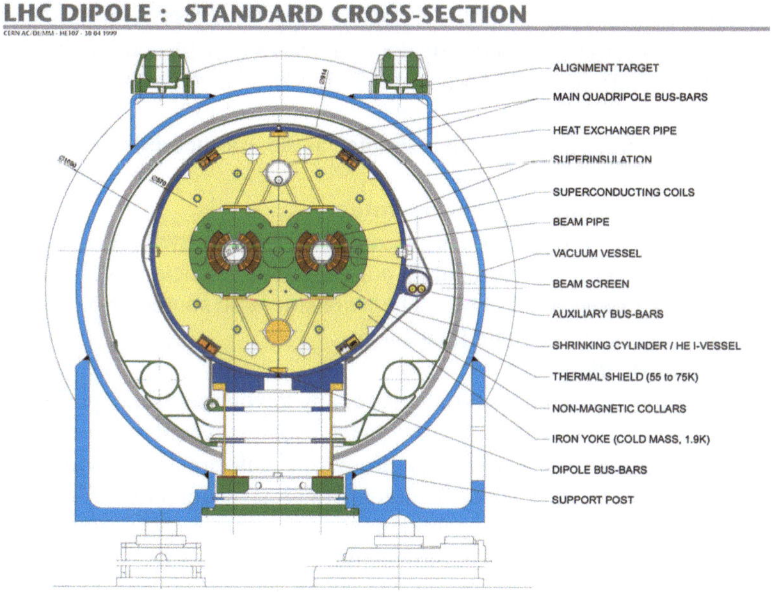

Fig. 2.3 Section of a LHC dipole magnet. The two separate rings are clearly visible and they are enclosed in the same cryostat

2.2 The LHCb Detector

The LHCb experiment is focused on the study of beauty and charm flavour physics at the LHC. With the large, measured $b\bar{b}$ production cross-section of 284 $\pm$ 20 $\pm$ 49 mb [6] at $\sqrt{s}$ = 7 TeV, it is the most abundant source of b-hadrons, which can hadronise into different species of B mesons, e.g B_d, B_u, B_s and B_c and other long lived Λ_b baryons. These unique properties allow studies of CP-violating phenomena, in both $B_{u,d}$ and B_s sectors, and rare decays with unprecedented precision. However, hadronic collisions result in a high-background environment, with events containing up to 300 charged tracks. The high luminosity delivered by the LHC also increases the chances of multiple interactions per event, making the reconstruction of b-hadron secondary vertices experimentally challenging.

The physics requirements needed for precise flavour physics studies, along with the experimental conditions, set stringent constraints on the performance of the LHCb detector. The key points can be summarized as follows:

- **Impact parameter and vertex resolution**
  An excellent vertex resolution is crucial in order to resolve secondary vertices coming from b-hadron and c-hadron decays with respect to the primary pp interactions. A performing proper time determination is particularly important for oscillation measurements, in particular for the fast oscillating B_s^0–$\bar{B}_s^0$ system. Furthermore, a good resolution provides a better distinction between displaced secondary tracks and prompt background tracks originating directly from the primary vertex. In the LHCb detector, vertexing is performed by the VErtex LOcator (VELO) subdetector, presented in Sect. 2.2.2, which is situated as close as possible to the interaction region.
- **Momentum and invariant mass resolution**
  A good resolution on the invariant mass of the reconstructed particles is needed to distinguish b-hadrons from random combinations of background tracks. Combinatorial background is expected to be a considerable source of background for physics analyses with hadronic final states.
- **Particle identification**
  Particle identification is an essential tool for any flavour physics experiment. Electrons and photons are identified via the electromagnetic calorimeter and muons are the only particles which can pass through the calorimeters leaving hits in the muon stations. As hadrons are difficult to distinguish using only information from the calorimeters, additional discrimination is needed to ensure a satisfactory pion, kaon and proton separation over a momentum range up to 100 GeV. Discrimination between charged hadrons is provided in LHCb by a Ring Imaging CHerenkov (RICH) system (Sect. 2.2.5), which uses three radiators with different optical properties.

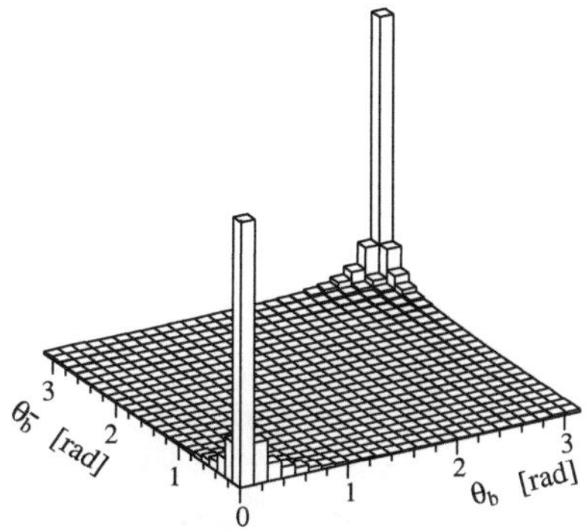

Fig. 2.4 Angular correlation between the b and $\bar{b}$ quarks produced in pp interactions (angles formed with respect to the beam axis). Events are simulated with the PYTHIA event generator [7]

- **Trigger**
  In a hadronic environment, b-hadron events are overwhelmed by uninteresting background events. A trigger system is required to select only useful collisions and reduce the total data bandwidth to an acceptable rate. The LHCb detector runs a multi-stage trigger, both hardware and software, which reduces the rate of visible interactions from about 10 MHz to about 3 kHz (Sect. 2.2.8). The system must be efficient and extremely flexible at the same time: it must select as many B final states as possible and adapt quickly to the variable LHC running conditions.

At the LHC, $b\bar{b}$ pairs are produced through the strong interaction and they carry the momentum of the partons involved in the hard collision. As the pp collision energy is much higher than the $b\bar{b}$ invariant mass, it is likely that there is considerable difference in the colliding partons' momenta. The $b\bar{b}$ pairs are therefore boosted in the direction of the most energetic parton, predominately at low angles in the laboratory system with respect to the beampipe. The trajectories of the resulting b-hadrons are highly correlated and concentrated in the forward/backwards regions of the colliding protons, as shown in Fig. 2.4. As a result of this forward production and the limited space available in the experimental area at IP8, the LHCb detector is designed as single-arm spectrometer with a coverage from approximately 10–300 (250) mrad in the horizontal bending (vertical non-bending) plane. This corresponds to a pseudorapidity[1] coverage of $1.9 < \eta < 4.9$. The LHCb detector is presented in Fig. 2.5. The interaction region is situated inside the VErtex Locator detector, at the left side of the plot, and it is displaced -11.25 m from the center of the cavern. The total length of the detector is about 20 m in the longitudinal direction (z-axis), 12 m in the horizontal direction (x-axis), and about 10 m in the vertical one (y-axis). The six subdetectors composing LHCb can be conveniently grouped as follows:

- **The tracking system** (Sects. 2.2.2, 2.2.3 and 2.2.4)
  composed by the VErtex LOcator (VELO), the LHCb magnet, the Tracker Turicensis (TT) and three tracking stations (T1,T2,T3).

[1] The pseudorapidity variable is defined as $\eta = -\ln\tan(\theta/2)$, with θ being the angle between the particle momentum and the beampipe.

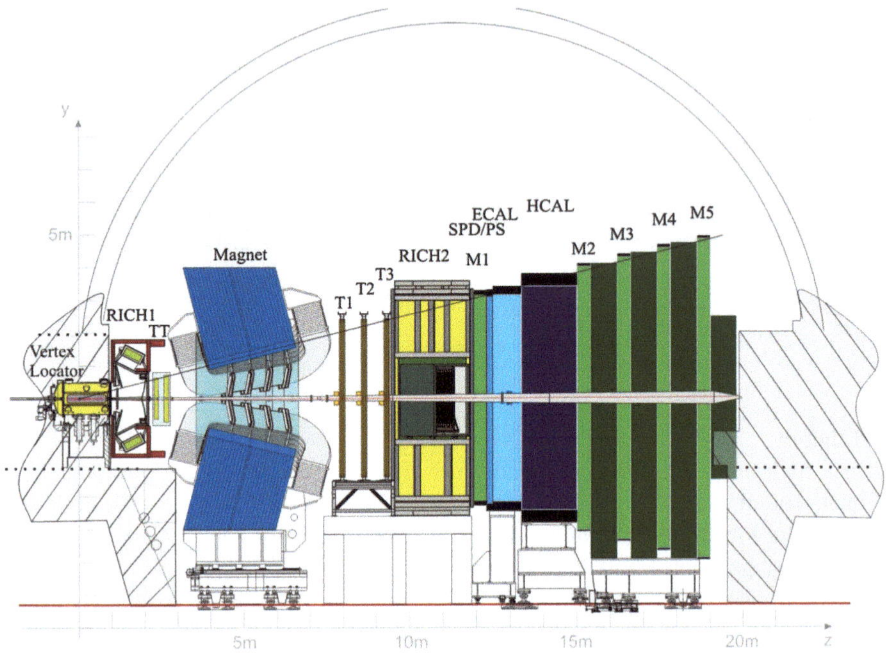

Fig. 2.5 Layout of LHCb detector on the transverse plane

- **The RICH detectors** (Sect. 2.2.5)
  composed by two separate systems RICH1 and RICH2.
- **The calorimeters** (Sect. 2.2.6)
  composed by the Scintillating Pad Detector (SPD), the Pre-Shower detector (PS), the Electromagnetic CALorimeter (ECAL) and the Hadronic CALorimeter (HCAL).
- **The muon stations** (Sect. 2.2.7)
  composed by five muon stations (M1,M2,M3,M4,M5).
- **The trigger system** (Sect. 2.2.8)
  composed by Level 0 (L0), High Level Trigger 1 (HLT1), High Level Trigger 2 (HLT2).

Each subsystem is described here briefly; detailed information on the LHCb detector can be found in its Technical Design Report [4].

2.2.1 LHCb Detector Running Conditions

This section presents a brief overview of the running conditions of the LHCb experiment in 2010 and 2011. The LHCb detector has performed well during both years, surpassing its target design luminosity and showing an excellent flexibility in adapting to the variable running conditions of the LHC. Two important parameters can be used to describe the operation of the LHCb detector: the instantaneous luminosity $\mathcal{L}$ and the average number of visible interactions per bunch crossing μ. In the LHCb interaction region, the luminosity is reduced and levelled in order to maintain a constant value during each fill. For the design luminosity $\mathcal{L} = 2 \times 10^{32}\text{cm}^{-2}\text{s}^{-1}$, the design μ is about 0.4–0.5, which means that the data collected with this experimental conditions should be dominated by events with one single primary vertex. In fact, a high value of μ indicates multiple pp interactions per event, resulting in high track multiplicity and increased background. The time needed to process every event by the trigger is also hugely affected as the number of track combinations scales factorially.

2010 Running Conditions

In 2010, the LHCb experiment collected about 37.7 pb^{-1}, with an average efficiency ($\mathcal{L}_{\text{recorded}}/\mathcal{L}_{\text{delivered}}$) of 89 %. Figure 2.6 shows the integrated luminosity recorded during 2010. Almost all data was collected in the last two weeks of running, reaching 80 % of the design instantaneous luminosity. However, the experimental conditions have been completely different from the design values, with an average value of $\mu > 2$ (see Fig. 2.7). As high values of μ greatly affected the event size and the online processing time, the trigger configurations and thresholds had to be changed regularly during 2010.

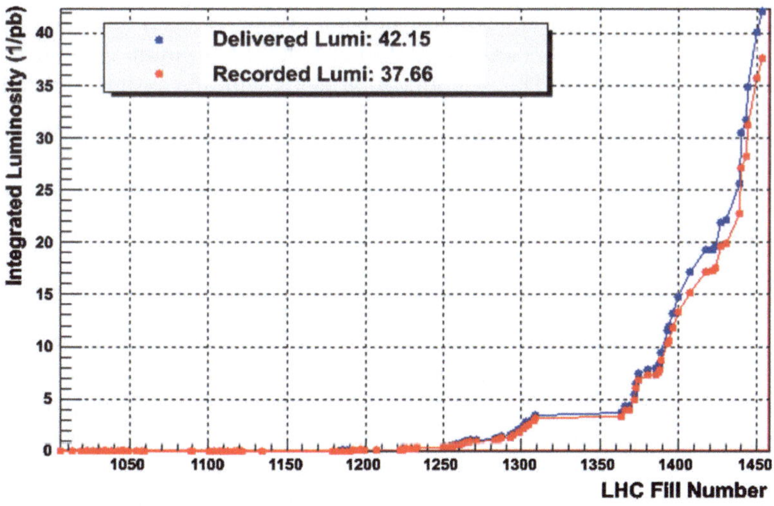

Fig. 2.6 Integrated luminosity versus LHC fill number delivered and collected in 2010

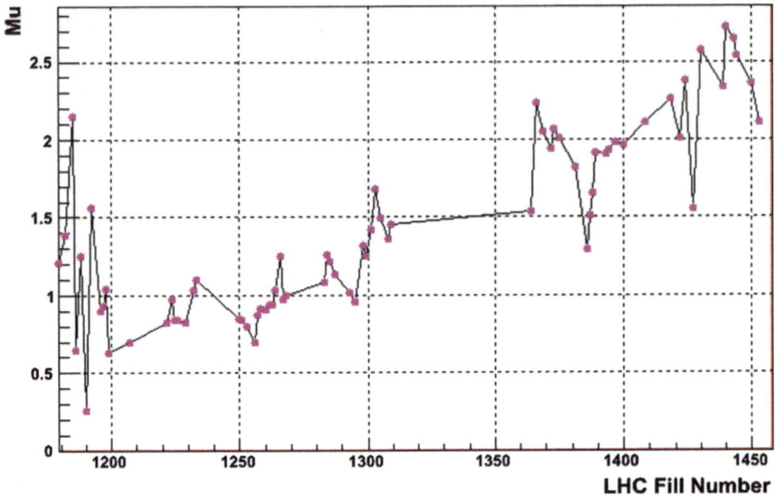

Fig. 2.7 Peak μ versus LHC fill number in 2010 (design value $\mu = 0.4$)

2011 Running Conditions

In 2011, data taking conditions were fairly stable at LHCb, recording more than 1 fb^{-1} of data at a constant pace, as shown in Fig. 2.8. The instantaneous luminosity was set to $\mathcal{L} = 3.5 \times 10^{32}\ \text{cm}^{-2}\text{s}^{-1}$, without a significant loss of data quality. The measured μ during 2011 is shown in Fig. 2.9 and was about three times above its design value (but lower than what observed in 2010). These stable experimental conditions resulted in more homogeneous trigger setup, as shown in Table 2.1, with only five different Trigger Configuration Keys (see Sect. 2.2.8) contributing to more than 90 % of the total collected luminosity. This simplification helps with the systematic studies of the trigger acceptance.

2.2.2 *VELO*

The VErtex LOcator (VELO) [8] provides precise measurements of the track position in the region close to the *pp* collisions. Primary vertices are reconstructed using the prompt tracks coming directly from the *pp* interactions and, as most of the long-lived *b* and *c*-hadrons decay inside the VELO instrumented volume, displaced tracks are used to identify and reconstruct their secondary vertices with good accuracy. Displaced tracks are also used as a trigger criterion in the HLT1 (Sect. 2.2.8). The resolution on the primary vertex is found to be, on average, about 42 μm (10 μm) on the *z* coordinate (x–y transverse plane) and depends on the number of tracks in the event. In order to achieve the desired resolution, the detector is located as close as

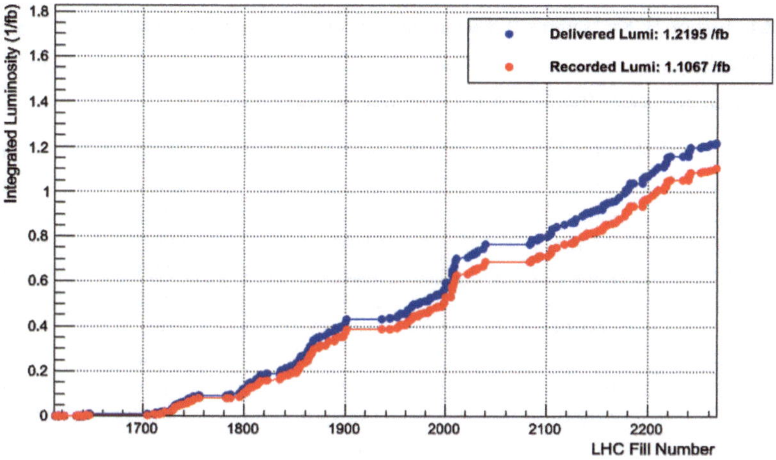

Fig. 2.8 Integrated luminosity versus LHC fill number delivered and collected in 2011

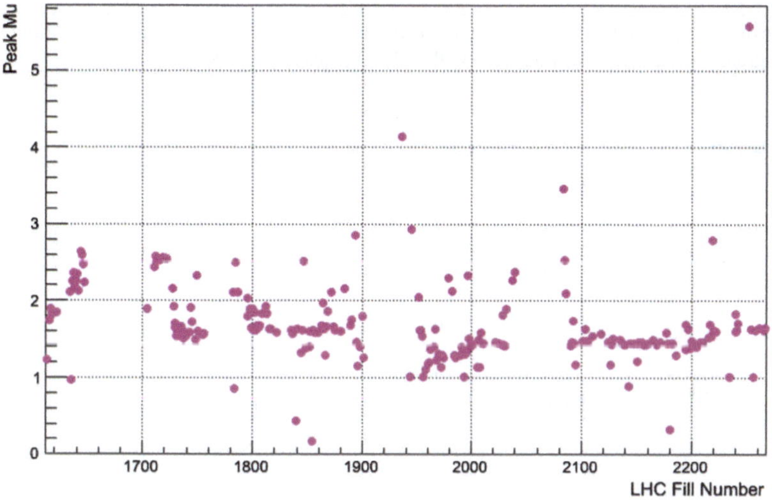

Fig. 2.9 Peak μ versus LHC fill number in 2011 (design value $\mu = 0.4$)

possible to the interaction region. The VELO layout is shown in Fig. 2.10: it consists of two sets of 21 modules positioned around the nominal interaction region perpendicularly to the beam axis. Each module provides an independent measurement of the r and ϕ cylindrical coordinates. The z coordinate is provided by the fixed z-position of the module itself.

Each station is formed by two semi-circular silicon detectors, with a sensitive region starting from 8.2 mm radially from the beam axis when the detector is operated in the *closed* position. The aperture increases during the LHC injection phase (*open* position) for safety reasons and the VELO is closed around the optimal beam

Table 2.1 List of TCKs used in 2011 data with the corresponding collected luminosities, split by magnet polarity

TCK	Magnet down	Magnet up	Total	(% rel.)
0x360032	3.36	-	3.36	(0.31)
0x480032	2.00	-	2.00	(0.18)
0x4A0033	2.09	-	2.09	(0.19)
0x561710	0.03	-	0.03	(0.00)
0x5A0032	28.45	38.63	67.08	(6.19)
0x5B0032	0.03	2.20	2.23	(0.21)
0x5D0033	-	2.25	2.25	(0.21)
0x6D0032	100.33	-	100.33	(9.25)
0x700034	1.14	-	1.14	(0.11)
0x710035	0.90	-	0.9	(0.08)
0x730035	61.83	134.35	196.18	(18.09)
0x740036		5.18	5.18	(0.48)
0x760037	191.64	107.11	298.75	(27.55)
0x790037	-	39.30	39.3	(3.62)
0x790038	209.43	153.97	363.4	(33.52)

Luminosities, expressed in pb^{-1}, are taken from the online LHCb $\mathcal{L}$ measurement. About 95 % of the integrated luminosity was collected with only five different TCKs

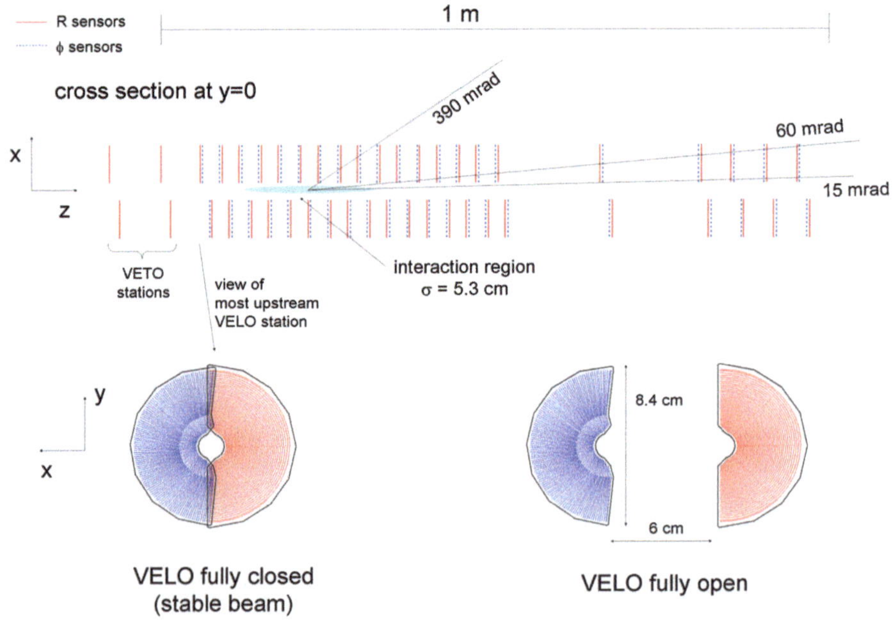

Fig. 2.10 *Top* layout in the x–z plane of the VELO sensors. The interaction region and angular acceptances are plotted. The acceptance region of the VELO is $1.6 < \eta < 4.9$. *Bottom* the front face of the first modules in the x–y transverse plane in closed (*left*) and open (*right*) position

x–y center only after stable beams are declared. For this purpose, the two halves of the detector are mounted on a controllable positioning system, located in a vessel that maintains the vacuum around the sensors, though separated from the machine vacuum by a thin walled corrugated aluminum sheet.

Two pile-up modules, containing only r-coordinate sensors, are located upstream of the VELO sensors. Those modules can be read at 40 MHz and they were originally designed to provide useful information to the L0 trigger (Sect. 2.2.8) in order to reject multiple interaction events. Given the changed experimental conditions (see Sect. 2.2.1), they are not currently used in the trigger.

2.2.3 Dipole Magnet

Charged tracks are bent in the horizontal plane with an intense magnetic field generated by a normal-conductive dipole magnet [9] weighting 1,600 tons. The LHCb magnet, shown in Fig. 2.11, provides an integrated magnetic field of 3.6 Tm. It is situated between the two RICH detectors and its magnetic field profile is designed in order to minimise the residual field inside the RICH1 boxes and the VELO, while creating a field as high as possible in the region before the tracking stations. In order to achieve the required momentum resolution, better than $\delta p/p \approx 0.5\,\%$ over a wide momentum range, the magnetic field must be known with a spacial precision up to a few millimeters.

Fig. 2.11 View of the LHCb dipole magnet

2.2.4 Tracking Stations

The coordinates on the transverse plane of the charged tracks are measured by five planar tracking stations located at different z-positions. The first two layers (TT) are situated before the magnet, while the latter three (T1, T2, T3) after. T1, T2 and T3 are composed of two different types of detector, the Inner Tracker (IT) and Outer Tracker (OT), and they are optimised in order to maximise the momentum measurement resolution. The layout of the LHCb tracking system is shown in Fig. 2.12.

Tracker Turicensis (TT)

The Tracker Turicensis (TT) is a 130×150 cm ($8.4\,\mathrm{m}^2$) planar tracking station and it is situated before the dipole magnet. It consists of four layers of silicon microstrips and covers the full LHCb acceptance. With a maximal strip occupancy of about 2.5 % in the region close to the beampipe, the expected spatial resolution is about 50 μm. These four instrumented layers are arranged in two pairs, separated about 27 cm in the

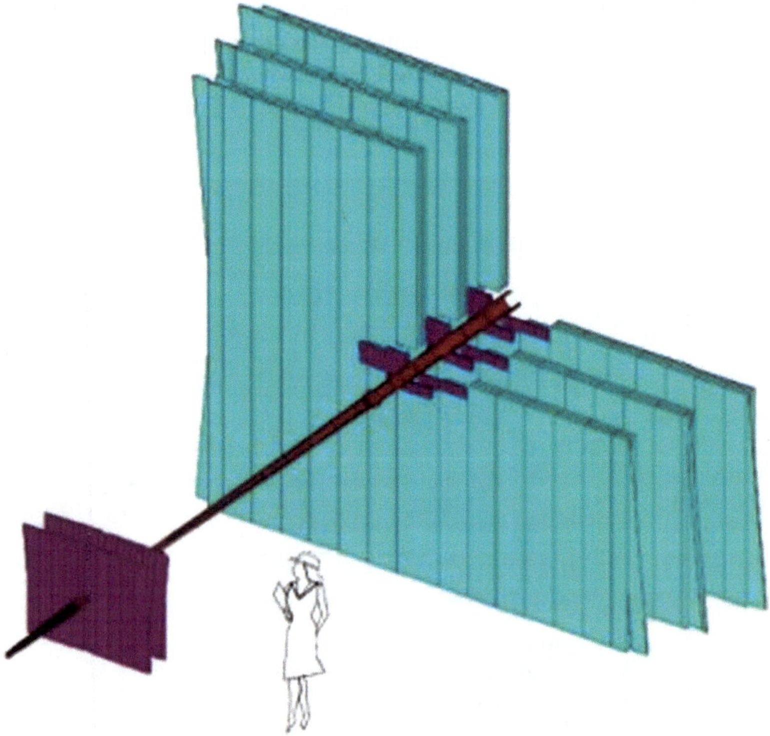

Fig. 2.12 Layout of the LHCb tracking system. The silicon microstrips of the TT and the IT are coloured in *purple*, the OT planes are shown in *blue*

z direction. The first and the last planes have vertical readout strips (x layer), while the second and the third have strips rotated by a small angle of $\pm 5^\circ$ respectively (u and v layers). This small stereo angle is chosen to maximise the momentum resolution, as charged particles are mainly bent in the horizontal plane. Information coming from the TT detector is important for the reconstruction of long-lived neutral K_s^0 mesons, that decay mostly outside the fiducial volume of the VELO. With a $c\tau \approx 2.7$ cm and relativistic factors γ of 10–100 at the LHC, K_s^0 mesons can travel up to several meters before decaying.

Inner Trackers (IT)

The Inner Tracker (IT) [10] composes the central part of the tracking stations situated after the dipole magnet. It covers the acceptance region close to the beampipe, where the particle flux is highest, and uses the same silicon strip technology as the TT. The cross-shaped layout of the detector (120 × 40 cm) covers only 1.3 % of the total sensitive area of each tracking plane, with a spatial resolution of 57 μm. However, as most of the tracks are produced in the forward/backward region along the beamline, about 20 % of the total number of charged tracks in each event is concentrated in this region. The Inner Tracker is shown in Fig. 2.13.

Outer Trackers (OT)

The Outer Tracker (OT) [11] is a drift-time detector providing acceptance coverage in the outer region of the tracking stations. Each module is composed of four layers of drift-tubes with diameters of 4.9 mm. The geometry of the four layers is identical to that of the TT detector, with the second and third layers rotated by a $\pm 5^\circ$ stereo

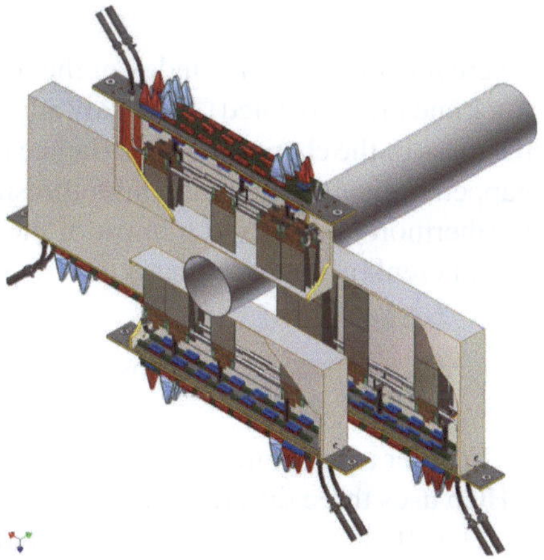

Fig. 2.13 A view of the IT

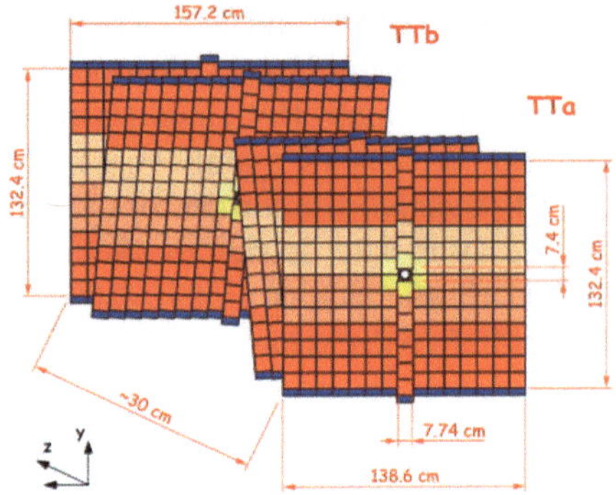

Fig. 2.14 A view of the OT

angle. Each drift-tube is filled with a mixture of Argon (70 %) and CO_2 (30 %), with drift times below 50 ns. The layout of the Outer Tracker is shown in Fig. 2.14.

2.2.5 RICH

Discrimination between charged hadrons is provided in the LHCb detector by two Ring Imaging CHerenkov detectors (RICH1 and RICH2) [12, 13]. Particle identification is particularly challenging at LHCb since mass discrimination is required over the broad momentum spectrum of the particle produced. RICH detectors use the light emitted by a charged track passing through a dielectric with a velocity v greater than the speed of light in that medium, the so-called Cherenkov effect [14]. Cherenkov photons are emitted into a cone of aperture angle θ_c around the direction of propagation, with:

$$\cos(\theta_c) = \frac{c}{nv} = \frac{1}{n\beta} \tag{2.1}$$

where n is the refractive index of the medium and $\beta = v/c$. If the momentum p of a track and the associated Cherenkov angle θ_c are measured together, a determination of the mass of the charged particle (hence its type) is possible. The Cherenkov effect can happen only above a momentum threshold depending on n and the type of particle. Furthermore, for a fixed n, most of the PID sensitivity is lost at high momentum as the Cherenkov angle saturates to

$$\cos(\theta_c^{\max}) \approx \frac{1}{n} \quad \text{for} \quad \beta \to 1 \tag{2.2}$$

In order to cover the momentum spectrum of the tracks produced in the collisions, LHCb uses three different radiators with different optical properties.

The RICH1 detector [15], presented in Fig. 2.15, is situated between the VELO and the dipole magnet. It is optimised to provide particle discrimination in the low

Fig. 2.15 A view of RICH1

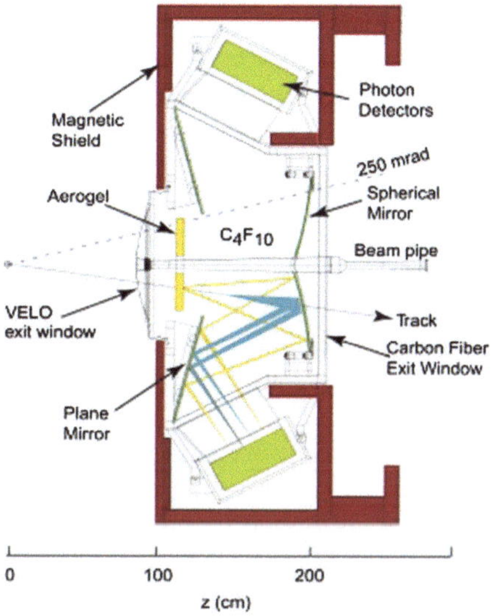

momentum range (1–60 GeV/c) with two different radiators: aerogel (n = 1.03) and C_4F_{10} gas (n = 1.0014). RICH1 covers the full LHCb angular acceptance.

The RICH2 detector [16], shown in Fig. 2.16, is located after the magnet and the tracking stations and before the calorimeters. Covering a limited angular acceptance of $\pm$ 15 to $\pm$ 200 mrad ($\pm$100mrad) in the horizontal (vertical) plane, it ensures particle identification for high momentum tracks above 15 GeV/c, up to about 100 GeV/c. The radiator used is CF_4 gas with a refractive index $n = 1.0005$.

In both RICH detectors, the Cherenkov light is deflected outside the geometrical acceptance of the detector via a set of spherical and flat mirrors and it is focused on arrays of Hybrid Photon Detectors (HPD). RICH1 (RICH2) hosts 196 (288) HPDs and each phototube is shielded from the residual magnetic field generated by the dipole.[2] HPD phototubes detect the light emitted and their information is used to reconstruct the Cherenkov cones originating from each charged track. Cones are seen as rings of hits on the phototube planes, as shown in Fig. 2.17.

2.2.6 Calorimeters

The LHCb calorimeter system [17] provides measurements of the energies and transverse positions of electrons, photons and hadrons. It plays a crucial role in the hadronic L0 trigger in order to recognise high E_T objects (Sect. 2.2.8). The classical layout of an electromagnetic calorimeter (ECAL) in front of a hadronic calorimeter

[2] Magnetic distortion corrections due to the residual magnetic field in the RICH vessels are taken into account during the offline reconstruction.

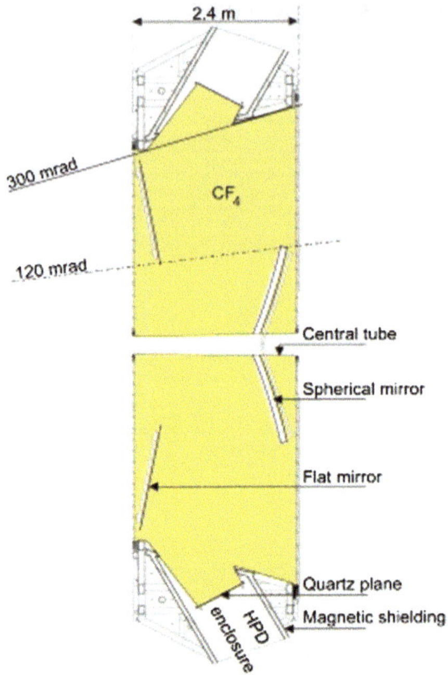

Fig. 2.16 A view of RICH2

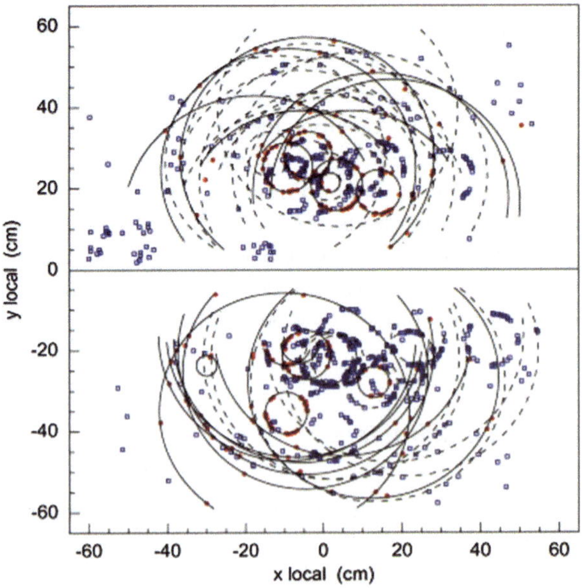

Fig. 2.17 Simulation of a LHCb event in RICH1. The Cherenkov light cones are reconstructed and fitted as rings (superimposed in the plot). The average number of Cherenkov photons emitted per track is small: about 8 in the aerogel, 30 in the C_4F_{10} and 22 in the CF_4

(HCAL) is adopted. Two additional detectors are placed in front of the ECAL to help electron discrimination from the high π^0 and γ backgrounds: the Scintillator PaD (SPD) and the Pre-Shower (PS) detectors. The calorimeters are situated between the first (M1) and the second (M2) muon stations. All calorimeter subdetectors are based on the same technology: segmented scintillators act as sensitive material and light is transported by wavelength-shifting (WLS) fibers to standard PhotoMultiplier Tubes (PMT).

SPD and PS

The Scintillator PaD (SPD) and Pre-Shower (PS) detectors are two almost identical layers of high-granularity scintillator pads. A lead converter of 15 mm, corresponding to 2.5 X_0, is located between the two scintillator planes and allows electrons to start showering before the ECAL. The SPD (PS) is situated before (after) the lead layer with a sensitive area of 7.6×6.2 m. Each scintillator tile is a square of side 3.9 cm (1,536 cells), 5.9 cm (1,792 cells) or 11.8 cm (2,688 cells) depending on the position with respect to the beamline. The geometry of a one-to-one projective correspondence with the ECAL segmentation is respected. The SPD is used to provide a good discrimination of electrons from photons in the L0 trigger and it is also used to provide a fast online estimation of the total number of charged tracks in the event. Furthermore, the background coming from charged pions is rejected analysing the longitudinal shower dispersion, measured confronting the SPD and PS hits.

ECAL

The Electromagnetic CALorimeter (ECAL) is shown in Figs. 2.18 and 2.19. It is a sampling calorimeter composed of 66 layers of lead absorber (2 mm) and scintillator

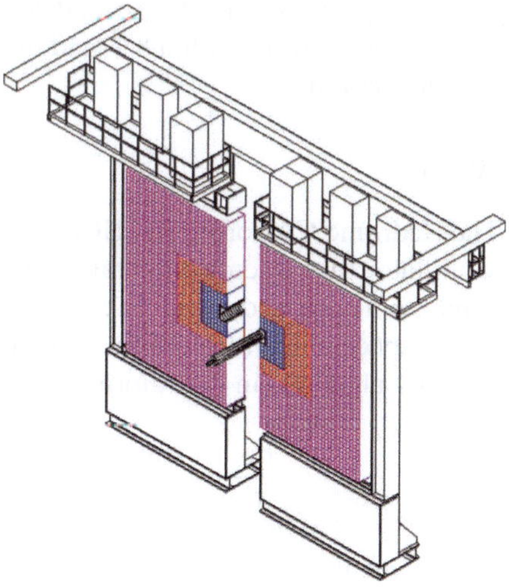

Fig. 2.18 Layout of the ECAL

Fig. 2.19 ECAL Photomultipliers

(4 mm), corresponding to about $25\,X_0$. This ensures a good confinement of all electromagnetic-interacting particles before the HCAL. The energy resolution of the ECAL is measured to be:

$$\frac{\sigma_E}{E} = \frac{8.5 - 9.5\,\%}{\sqrt{E}} \oplus 0.8\,\% \quad \text{with } E \text{ in GeV} \tag{2.3}$$

depending on the module type, where the first term is the statistical error on the energy deposition and the second a constant term. ECAL module responses are found to be uniform within 8 % and the measured parameters are consistent with the design performance.

HCAL

The Hadronic CALorimeter (HCAL) consists of layers of iron and scintillator, with scintillator tiles orientated parallel to the z-axis. With a total dimension of 8.4 m in height, 6.8 m in width and 1.65 m in length, the detector thickness is only 5.6 λ_{int} due to space constraints in the experimental area. Therefore, hadronic showers are not fully absorbed in the calorimeter and the energy resolution is measured as:

$$\frac{\sigma_E}{E} = \frac{(69 \pm 5)\,\%}{\sqrt{E}} \oplus (9 \pm 2)\,\% \quad \text{with } E \text{ in GeV} \tag{2.4}$$

in agreement with the design values.

2.2.7 Muon Stations

The muon stations [18] are located at the end of the detector and provide efficient muon identification and triggering. The muon system is composed of five station (1,380 chambers in total) of rectangular shape: the first station M1 is located upstream the calorimeters and provides an improved p_{T} measurement for the trigger. The other stations (M2 to M5) are placed after the HCAL and are sandwiched with 80 cm thick iron planes. The iron absorbers stop the hadrons punching through the hadronic calorimeter and select only penetrating muons, with a minimum momentum of 6 GeV/c required to reach M5. The layout of the muon system is sketched in Fig. 2.20. The total angular acceptances are 20–306 mrad in the bending plane and 16–258 mrad in the non-bending one. Multi-Wire Proportional Chambers (MWPC) are used in all the stations, except in the inner region of M1 where triple-GEM (Gas Electron Multiplier) detectors have been preferred due to their better radiation hardness (in presence of high particle rates, GEMs can operate at higher gas gains compared to MWPCs, while decreasing the risk of discharges). The granularity of the stations is chosen according to the expected particle flux in that region. Stations M4 and M5 have limited spatial resolution and their main purpose is the identification of high p_{T} penetrating particles.

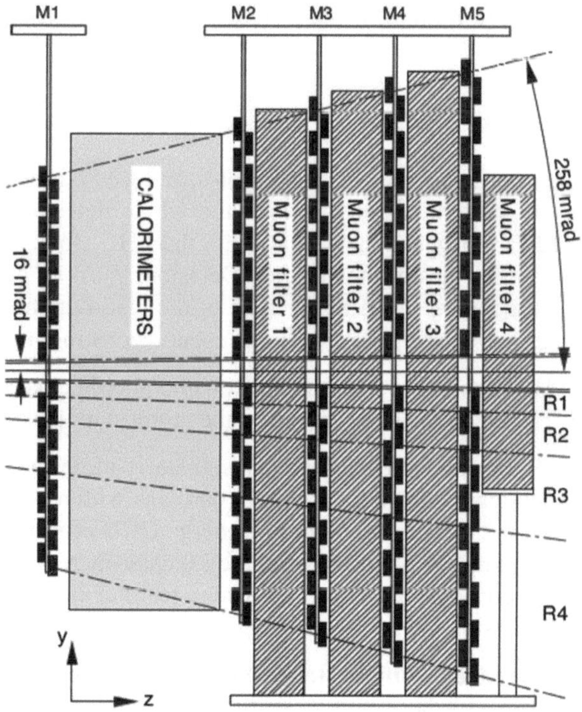

Fig. 2.20 Layout of the muon chambers and iron absorbers

2.2.8 *Trigger*

The LHCb detector runs a three-stage trigger [19, 20] in order to reduce the rate of visible interactions to a value of about 2–3 kHz. This represents the experimental bandwidth limit for writing events to offline storage. The LHCb trigger must be able to select the largest possible fraction of interesting collisions in a high background environment, with b and c-hadrons reconstructed into a broad range of final states. In 2010 and 2011, the LHCb experiment had to cope with running conditions very different from the design values (see Sect. 2.2.1). With visible pp collision rates of 12–15 MHz, dominated by events with multiple primary vertices, the event size has increased considerably. Therefore, the trigger working conditions had to be adjusted accordingly due to a higher-than-designed online processing time. The total output of the trigger in 2011 has been about 3 kHz, exceeding the designed value of a factor of 50 %. In 2012, the trigger output rate has been increased to 4.5 kHz.

The trigger system is divided in two levels: the Level-0 (L0) trigger and the High Level Trigger (HLT). The L0 is a hardware trigger which operates synchronously with the LHC clock, while the HLT is fully implemented within software and processes the output of the L0 on the Event Filter Farm (EFF). The EFF is composed of more than 20,000 CPU cores, which process events in parallel. The HLT is divided into two stages: the HLT1 performs only a partial inclusive reconstruction of the event and once the rate is reduced by a factor of about 20, the HLT2 performs a *almost-full* online event reconstruction and selects events for storage.

L0 Trigger

The first trigger stage, the Level-0, reduces the event rate to about 1 MHz. This is the maximum rate at which the whole detector can be read out. The L0 is a hardware trigger synchronous with the LHC clock and it is integrated in the front-end electronics of the detector. Decisions must be provided with a latency of 4 μs corresponding to the buffer time of the read-out chips. The L0 trigger exploits the large p_{T} objects produced in the event reconstructing:

- The highest E_T hadron, electron and photon clusters in the calorimeters.
- The two highest p_{T} muons in the muon stations.

These are good experimental signatures of heavy particle decays, e.g. b-hadrons, which tend to produce daughter particles with a large transverse momentum compared to the background of low-energy QCD processes. The L0 trigger is composed of three independent systems: the L0 calorimeter triggers, the L0 muon trigger and the pile-up system. Those are read in parallel and a final decision is provided by the L0 Decision Unit (L0 DU) from the logical *or* of their outputs:

- **The L0 calorimeter trigger** forms clusters of 2×2 calorimeter cells and selects those with the highest E_T. Clusters are tagged as electron (`L0Electron`), photon (`L0Photon`) or hadron (`L0Hadron`) depending on the energy deposits in the

SPD/PS, ECAL and HCAL. The HCAL is also used to reject events triggered by halo muons, while the total number of hits in the SPD is used to provide a fast, hardware estimation of the charged track multiplicity in the event.[3]

- **The L0 muon trigger** [21] looks for muon tracks with a large p_T (`L0Muon` and `L0DiMuon` lines). A fast, stand-alone procedure is used to search for hits in the muon stations compatible with a straight line pointing towards the interaction point. The highest-p_T muon segments are used for the decision.
- **The pile-up system** uses the two pile-up modules situated upstream the VELO (Sect. 2.2.2). The system was originally designed to provide a fast decision on multiple interaction events and it is currently used only to trigger beam-gas interactions.

HLT1 Trigger

The HLT1 trigger reconstructs particles in the VELO and gives a first estimation of the position of the primary vertices (PVs) in the event. HLT1 should reduce the rate to a sufficiently low level, approximately 50 kHz, to allow a complete reconstruction of the selected events in the HLT2. Several trigger decision lines are run in parallel, with a total decision time per event of order of 10–15 ms:

- **The single track trigger** [22] is used to select the decays of long-lived b and c hadrons. Its idea is to search for a single displaced high momentum track in the event. Three-dimensional VELO tracks act as a seed for reconstruction and no confirmation of the L0 trigger decision is required at this stage. A good quality track with a large impact parameter (IP) with respect to the primary vertices (PV) in the event is required to trigger the event.
- **Muon HLT1 triggers** [23] are executed only if the event has been triggered by `L0Muon` *or* `L0DiMuon` lines. All the VELO tracks are extrapolated and matched with the muon segments used in the L0 decision (there is no confirmation of the L0 candidates). This provides lifetime unbiased muon track samples. Single and di-muon lines are run depending on the requirements applied on the momentum of the particles and the invariant mass of the composed objects.
- **Electron and photon triggers** exploit the information coming from the ECAL.

HLT2 Trigger

The HLT2 trigger performs an *almost-full* pattern recognition of the events passing the HLT1 selections. Therefore, the HLT2 trigger lines use selection criteria which are as close as possible to the offline selections. The HLT2 accesses the full event information and it is able to look for secondary vertices and composed particles.

[3] In 2010, events with $n_{\rm hits}^{\rm SPD} > 900$ were rejected to keep the OT occupancy at about 20 %. In 2011, a looser cut of $n_{\rm hits}^{\rm SPD} > 600$ was chosen due to improved running conditions for the same to give better signal efficiency.

Global Event Cuts (GEC), such as the total number of reconstructed tracks, are also used to reject high multiplicity events which would require a too long processing time. A set of inclusive and exclusive selections are applied at this stage in order to take into account the largest possible number of final states.

The HLT2 muon lines are similar to those run in the HLT1, but with a more refined and complete track reconstruction. The n-body topological lines [24], instead, are designed to trigger on decays of $B \to$ anything, exploiting common decay properties of b-hadrons. In particular, $B \to DX$ decays are selected by the `Hlt2Topo2`, `Hlt2Topo3` and `Hlt2Topo4` lines, where the D meson decays into a two, three or four-body final state, respectively. Those lines are based on a fast Boosted Decision Tree (BDT) selection, which provides a good background rejection[4] and a high signal efficiency.

The HLT2-selected events are saved on permanent storage. All the created signal trigger candidates are stored in the raw event as well. This allows a direct comparison between the trigger objects that have fired a given line and the offline reconstructed signal candidates (see Sect. 3.6). In 2011, the HLT2 bandwidth was roughly shared equally among the topological, muon and charm triggers. Those lines account for almost all the total output bandwidth of about 3 kHz, as shown in Table 2.2 [25]. The rest is used by several exclusive lines, e.g. the radiative decay lines [26], and inclusive lines such as prompt J/ψ and ϕ triggers [27].

Trigger Configuration Keys (TCK)

Each trigger configuration used in data-taking is recorded and labeled via a unique identifier, called Trigger Configuration Key (TCK). Each TCK defines unambiguously the sequence of algorithms and trigger lines that have been run in the L0, HLT1

Table 2.2 Typical HLT2 trigger rates in 2011

HLT2 lines (TCK = 0x00760037, 3/8/2011)	Rate (Hz)
OR of Hlt2 *Topo decisions	1309.91
OR of Hlt2 *Charm decisions	822.92
OR of Hlt2 *SingleMuon decisions	480.91
OR of Hlt2 *DiMuon decisions	330.75
OR of Hlt2 *Muon decisions	819.34
OR of Hlt2 *DiProton decisions	45.83
OR of Hlt2 *SingleElectron decisions	1.04
OR of Hlt2 *Electron decisions	54.88
OR of Hlt2 *Gamma decisions	39.21
OR of Hlt2 *Phi decisions	77.56
OR of Hlt2 *ALL decisions	3012.19

Run: 97,871 (L0 800 kHz, $\mu = 1.4$). In the analysis presented in this thesis, all events come through the top line of the table

[4] The background rejection obtained with the BDT topological lines, used in 2011, is a factor three better than that obtained with the cut-based approach used in 2010.

and HLT2, including their configuration parameters and thresholds. TCKs are represented by a 32-bit number, with the first 16 bits reserved for the L0 configuration and the second 16 bits for the HLT. If the same lines are run, but with different thresholds, a different TCK number is associated to the new configuration. This labeling system was adopted to allow a quick configuration of the event filter farm and an easy inspection of the trigger properties.

2.2.9 Track Reconstruction

Charged tracks are refitted using the hit information coming from the VELO, TT, IT and OT [28, 29]. Depending on the properties of their trajectories, tracks can be divided into different classes, as sketched in Fig. 2.21:

- **Long tracks**
  They transverse the full tracking system, from the VELO to the last tracking station, leaving hits in all the subdetectors. These tracks have the most precisely determined momentum. All the tracks used for the analysis presented in this thesis are of this type.
- **Upstream tracks**
  They transverse only the VELO and the TT stations. These low momentum tracks are bent out of the detector acceptance by the magnetic field and they don't reach the tracking stations T1, T2 and T3. The momentum resolution on these tracks is usually poor. However, if their velocities are above the Cherenkov radiation threshold in RICH1 (Sect. 2.2.5), they can be used for its calibration.
- **Downstream tracks**
  They originate outside the VELO and they have hits only in the TT and T stations. Those highly displaced tracks are usually produced in the decay of long-lived

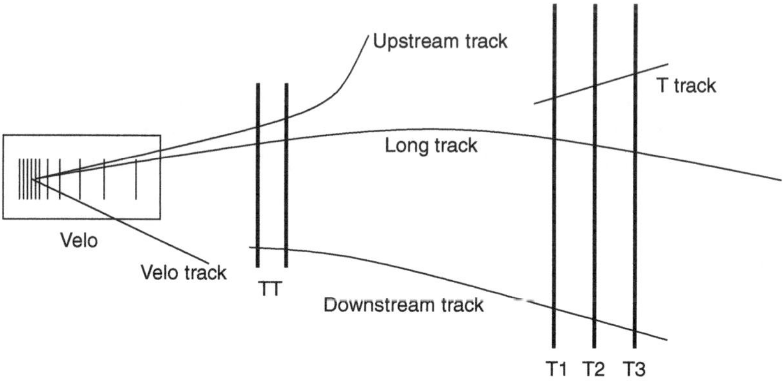

Fig. 2.21 A schematic view of the different types of charged tracks in the LHCb detector. Tracks are classified according to the detectors involved in their reconstruction

particles which decay outside the fiducial volume of the VELO, e.g. K_s^0 and Λ^0 hadrons.

- **VELO tracks**
  They exit the detector angular acceptance before the TT and have hits in the VELO detector only. Usually produced at low rapidity, they are useful for the reconstruction of the primary interaction vertices.
- **T tracks**
  They are only observed in the T stations. Produced in secondary decays, they cross the RICH2 detector and they can be used for its calibration.

An initial *seed* track is needed for the full track reconstruction. Straight-line segments of almost aligned hits are searched for in the VELO and T stations, where the magnetic field is low and a linear fit is a good approximation. After the seed tracks have been found, the hits from different detectors are matched and the trajectories refitted with a Kalman filter fit. Multiple scattering and $\mathrm{d}E/\mathrm{d}x$ energy loss are taken into account as well. The χ^2_{track} of the fit is a good indicator of the track quality and tracks with a poor χ^2_{track} can be rejected during the preliminary reconstruction of the signal candidates (see Sect. 3.4). Furthermore, a clone algorithm can check if two or more tracks have hits in common (duplicated segments) and remove any duplicated track [29].

Two important additional parameters can be introduced in order to evaluate the total performance of the track reconstruction:

- **Reconstruction efficiency**
  The reconstruction efficiency is the fraction of reconstructible tracks which are correctly reconstructed by the tracking algorithm. To be considered reconstructible, a track must lay inside the subdetectors' acceptance and give enough hits to be detected.
- **Ghost rate**
  A track is defined *ghost* if it is not matched with a charged particle in the event. These *fake* tracks are reconstructed from uncorrelated real/noise hits and have usually low momentum. The ghost rate is defined as the average fraction of ghost tracks in each event.

The reconstruction efficiency depends on several factors, e.g. the momentum spectrum of the tracks considered and the track multiplicity of the event. Using J/ψ decay products, the overall efficiency for long tracks is measured to be $95.4 \pm 0.7\,\%$ for 2010 data and $96.8 \pm 0.1\,\%$ for 2011 data [30]. The ghost rate depends on the algorithm used and it is found to be negligible for the $B \to [hh]_D h$ decays studied in this analysis.

2.2.10 Particle Identification (PID)

Particle identification of charged tracks is provided combining information coming from the RICH detectors, the calorimeters and the muon stations. For each track, all

PID contributions are merged into a log-likelihood difference ($\Delta LL_{h/\pi}$) between a given particle hypothesis h (with $h = K, p, \mu$ or e) and the pion hypothesis.[5] The global likelihood function is defined as:

$$\Delta LL_{h/\pi} = \ln L(h) - \ln L(\pi) = \ln\left[\frac{L(h)}{L(\pi)}\right] \tag{2.5}$$

where $L(h)$ is the likelihood of a given track of being a h hadron. It should be noted that the absolute value of the likelihood itself is not a useful quantity, as the scale changes in each event. The $\Delta LL_{h/k}$ between any h and k particle hypothesis can thus be derived trivially as follows:

$$\Delta LL_{h/k} = \Delta LL_{h/\pi} - \Delta LL_{k/\pi} = \ln L(h) - \ln L(k) = \ln\left[\frac{L(h)}{L(k)}\right] \tag{2.6}$$

The global likelihood hypothesis for each particle type is formed multiplying the different subdetector contributions:

$$L(K) = L^{\mathrm{RICH}}(K) \cdot L^{\mathrm{CALO}}(\mathrm{non}\; e) \cdot L^{\mathrm{MUON}}(\mathrm{non}\; \mu) \tag{2.7}$$

$$L(\pi) = L^{\mathrm{RICH}}(\pi) \cdot L^{\mathrm{CALO}}(\mathrm{non}\; e) \cdot L^{\mathrm{MUON}}(\mathrm{non}\; \mu) \tag{2.8}$$

$$L(\mu) = L^{\mathrm{RICH}}(\mu) \cdot L^{\mathrm{CALO}}(\mathrm{non}\; e) \cdot L^{\mathrm{MUON}}(\mu) \tag{2.9}$$

$$L(e) = L^{\mathrm{RICH}}(e) \cdot L^{\mathrm{CALO}}(e) \cdot L^{\mathrm{MUON}}(\mathrm{non}\; \mu) \tag{2.10}$$

Hadron Identification

Hadron discrimination derives mostly from the two RICH detectors. The circular rings observed on the phototube planes are compared with the pattern expected under a given particle hypothesis and the likelihood $L^{\mathrm{RICH}}(h)$ is calculated from this comparison [31, 32]. The performance of particle identification is assessed on data using a series of independent, high statistics calibration modes. For example, $D^{*+} \to D^0(K^-\pi^+)\pi^+$ decays provide a large unbiased data sample of pions and kaons (for more details on the calibration procedure, see Sect. 3.11). The RICH system is designed primarily to provide an excellent K/π discrimination over a broad momentum range, as shown in Fig. 2.22 [33]. The discrimination power depends on several factors, e.g the momentum p of the particle considered.

Muon Identification

Muon identification can use information coming from the muon chambers for track with $p > 3$ GeV/c. This is the minimum momentum required to pass the

[5] The pion hypothesis is used as an arbitrary *reference* mass hypothesis.

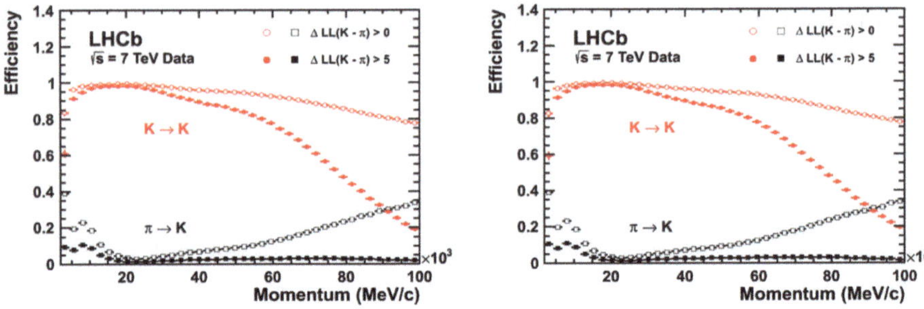

Fig. 2.22 RICH performance plots for the 2011 1.0 fb^{-1} dataset. *Left* magnet down polarity. *Right* magnet up polarity. $K \to K$ efficiency and $\pi \to K$ misidentification rate are plotted with respect to p and for two different$\Delta LL_{K/\pi}$ cuts

Table 2.3 Muon stations (from 1 to 5) required to trigger the muon decision as a function of momentum range

Momentum range	Muon stations
$3\,\text{GeV}/c < p < 6\,\text{GeV}/c$	M2 + M3
$6\,\text{GeV}/c < p < 10\,\text{GeV}/c$	M2 + M3 + (M4 or M5)
$p > 10\,\text{GeV}/c$	M2 + M3 + M4 + M5

The higher the momentum the stronger the requirement

calorimeters and reach the M2 station. Tracks are *flagged* as muons if they are associated to a certain number of hits in the five muon stations which depends on their momentum. The complete procedure is described in [34–36] and it is summarized in Table 2.3. A hit is considered consistent with a track if it is found within a Field Of Interest (FOI) defined around the track extrapolation in the muon chambers. The spatial size of the FOI is parametrized as a function of momentum and the region crossed. Furthermore, for the tracks passing the muon requirement, a hypothesis test is performed by evaluating for each track the compatibility with the muon and non-muon hypotheses. The two obtained likelihoods $L^{\text{MUON}}(\mu)$ and $L^{\text{MUON}}(\text{non}\ \mu)$ can be combined with the corresponding RICH likelihood $L^{\text{RICH}}(\mu)$ in order to provide a global $\Delta LL_{\mu/\pi}$ PID variable.

Electron and Neutral Particle Identification

Electron and neutral particle identification exploits the energy deposited in the ECAL to build a likelihood $L^{\text{CALO}}(h)$. As the analysis described in this thesis does not involve electrons or neutrals directly, PID at LHCb for those type of particles is not described here. More details on particle identification with the LHCb calorimeters can be found in [37], while γ and π^0 reconstruction is discussed in detail in [38].

2.2.11 LHCb Software Framework

The LHCb software environment is based upon the GAUDI framework [39]. GAUDI is a full Object Oriented framework used for all the software packages and applications of the LHCb experiment. The full Monte Carlo simulation, reconstruction and data analysis is developed in the same common environment. The main LHCb software tools are:

- GAUSS
  Simulation of LHCb events is handled by the GAUSS application [40]. GAUSS incorporates both event generation and full detector simulation tasks. *pp* primary interactions are generated with a tuned version of PYTHIA [7], while the decays of the intermediate particles are handled with the EVTGEN package [41]. After generation, the interaction of particles in the detector is described with GEANT4 [42]. Acceptance cuts can be applied at the generator level to increase the production efficiency.
- BOOLE
  The simulation of each subdetector response to the output of GAUSS is performed with a separate application: the BOOLE software package [43]. BOOLE simulates the digitization of the energy deposited in the LHCb detector active material. The L0 hardware trigger is emulated at this stage as well. The simulation of the detector electronics takes into account possible overlap between different *pp* events, the so-called spillover. The output format of BOOLE is exactly that used in real raw data files, therefore, simulated and real data files can be reconstructed with the same software chain.
- MOORE
  The Moore package [44] is used to run the HLT1 and HLT2 triggers in both the online system and offline simulated events. The trigger settings are defined via a unique Trigger Configuration Key (Sect. 2.2.8). Each TCK must be processed with a specific version of the MOORE software (the same that was run in the online event filter farm at the time data was taken) to ensure a correct reproduction of the trigger performance.
- BRUNEL
  The raw detector hits are read by BRUNEL [45] in order to reconstruct basic physics objects, e.g charged tracks, calorimeter clusters and Cherenkov rings. BRUNEL accesses the LHCb condition database in order to retrieve the experimental conditions of the detector at the time each event was taken. Physical objects are saved into .dst files for data analysis.
- DAVINCI
  DAVINCI [46] represents the last stage of the reconstruction. It is used to create composed objects, e.g. the composed particles in a decay chain, and perform cadidate selections. Several tools are available to construct and save useful quantities for offline analyses. Once candidates are selected, they can be stored in different file formats, ready for further analysis and visualization.

References

1. L. Evans, P. Bryant, LHC machine. J. Instrum. **3**(08), S08001 (2008)
2. The ATLAS Collaboration and others, The ATLAS experiment at the CERN large hadron collider. J. Instrum. **3**(08), S08003 (2008)
3. The CMS Collaboration and others, The CMS experiment at the CERN LHC. J. Instrum. **3**(08). S08004 (2008)
4. A. Alves et al., The LHCb detector at the LHC, J. Instrum. **3**, S08005 (2008) , Also published by CERN Geneva in 2010
5. The ALICE Collaboration and others, The ALICE experiment at the CERN LHC. J. Instrum. **3**(08), S08002 (2008)
6. LHCb Collaboration, R. Aaij et al., Measurement of $\sigma(pp \rightarrow b\bar{b}X)$ at $\sqrt{s} = 7$ TeV in the forward region, Phys. Lett. B**694**, 209 (2010), http://arXiv:1009.2731
7. S. Torbjörn, M. Stephen, S. Peter, PYTHIA 6.4 physics and manual, JHEP **05**, 026 (2006), arXiv:hep-ph/0603175
8. P.R. Barbosa-Marinho et al., LHCb VELO, VErtex LOcator: Technical Design Report. Technical Design Report LHCb. CERN, Geneva, 2001
9. S. Amato et al., LHCb magnet: Technical Design Report Technical Design Report LHCb. CERN, Geneva, 2000
10. P.R. Barbosa-Marinho et al., LHCb inner tracker: Technical Design Report. Technical Design Report LHCb. CERN, Geneva, 2002
11. P.R. Barbosa-Marinho et al., LHCb outer tracker: Technical Design Report. Technical Design Report LHCb. CERN, Geneva, 2001
12. S. Amato et al., LHCb RICH: Technical Design Report. Technical Design Report LHCb. CERN, Geneva, 2000
13. A. Pickford, The RICH system of the LHCb experiment: status and performance, Nucl. Instrum. Methods Phys. Res. A **604**(12), 297 (2009)
14. P.A. Cherenkov, Visible emission of clean liquids by action of γ radiation. Doklady Akademii Nauk SSSR **3**, 451 (1934)
15. N. Brook et al., LHCb RICH1 Engineering Design Review Report, Technical Report. LHCb-2004-121. CERN-LHCb-2004-121, CERN, Geneva, Oct 2005
16. E. Albrecht et al., LHCb RICH 2 engineering design review report, Technical Report. LHCb-2002-009, CERN, Geneva, May 2002
17. S. Amato et al., LHCb calorimeters: Technical Design Report. Technical Design Report LHCb. CERN, Geneva, 2000
18. P.R Barbosa-Marinho et al., LHCb muon system: Technical Design Report. Technical Design Report LHCb. CERN, Geneva, 2001
19. R. Antunes-Nobrega, et al., LHCb trigger system: Technical Design Report. Technical Design Report LHCb. CERN, Geneva, 2003. revised version number 1 submitted on 2003–09-24 12:12:22
20. H. Ruiz, The LHCb trigger: algorithms and performance, Nucl. Instrum. Methods Phys. Res., Sect. A **623**(1), 525 (2010)
21. E. Aslanides et al., The Level-0 muon trigger for the LHCb experiment, Nucl. Instrum. Methods Phys. Res., Sect. A **579**(3), 989 (2007)
22. Gligorov, V. Vladimir, A single track HLT1 trigger, Technical Report. LHCb-PUB-2011-003. CERN-LHCb-PUB-2011-003. LHCb-INT-2010-053, CERN, Geneva, Jan 2011
23. A. Roel, A. Johannes, Muon triggers in the High Level Trigger of LHCb, Tech. Rep. LHCb-PUB-2011-017. CERN-LHCb-PUB-2011-017, CERN, Geneva, Sep 2011
24. M. Williams, V. Gligorov, C. Thomas, H. Dijkstra, J. Nardulli, P. Spradlin , The HLT2 Topological Lines, Technical Report. LHCb-PUB-2011-002. CERN-LHCb-PUB-2011-002, CERN, Geneva, Jan 2011
25. https://twiki.cern.ch/twiki/bin/view/LHCb/Hlt2Rates
26. F. Soomro, V. Belyaev, HLT2 exclusive selections for $B_s \rightarrow \phi\gamma$ and $B_d \rightarrow K^*\gamma$, Technical Report. LHCb-PUB-2010-007. CERN-LHCb-PUB-2010-007, CERN, Geneva, Feb 2010

27. M. Lieng, An Inclusive phi Stream for the LHCb High Level Trigger, DC06 Analysis, Technical Report. LHCb-2009-010. CERN-LHCb-2009-010, CERN, Geneva, Jan 2009
28. O. Callot, S. Hansmann-Menzemer, The Forward Tracking: Algorithm and Performance Studies, Technical Report. LHCb-2007-015. CERN-LHCb-2007-015, CERN, Geneva, May 2007
29. https://twiki.cern.ch/twiki/bin/view/LHCb/LHCbTrackingStrategies#TrackTypes
30. A. Jaeger et al., Technical Report. LHCb-PUB-2011-025. CERN-LHCb-PUB-2011-025, CERN, Geneva, Apr 2012
31. R. Forty, RICH pattern recognition for LHCb, Nucl. Instrum. Methods Phys. Res., Sect. A **433**(12), 257 (1999)
32. M. Benayoun, C. Jones, RICH Reconstruction and Particle Identification Using Ring Fit Methods: Application to the RICH2 Detector, Technical Report. LHCb-2004-057. CERN-LHCb-2004-057, CERN, Geneva, Jan 2005
33. https://lbtwiki.cern.ch/bin/view/RICH/RichPIDPerformancePlots
34. F. Archilli et al., Muon Identification performance at LHCb with the 2010 data, Technical Report. LHCb-INT-2011-048. CERN-LHCb-INT-2011-048, CERN, Geneva, Oct 2011
35. K. Akiba, et al., Results on Muon identification efficiency with 2011 data at LHCb, Technical Report. LHCb-INT-2011-045. CERN-LHCb-INT-2011-045, CERN, Geneva, Oct 2011
36. E. Polycarpo Macedo et al., Performance of the Muon Identification in LHCb with 2011 data, Technical Report. LHCb-INT-2012-016. CERN-LHCb-INT-2012-016, CERN, Geneva, Jun 2012
37. H. Terrier, I. Belyaev, Particle identification with LHCb calorimeters, Technical Report. LHCb-2003-092, CERN, Geneva, Sep 2003
38. O. Deschamps, X. Machefert, M.H Schune, G. Pakhlova, I. Belyaev, Photon and neutral pion reconstruction, Technical Report. LHCb-2003-091, CERN, Geneva, Sep 2003
39. G. Barrand et al., GAUDI A software architecture and framework for building HEP data processing applications, Comput. Phys. Commun. **140**(12), 45 (2001)
40. I. Belyaev, P. Charpentier, S. Easo, P. Mato, J. Palacios et al., Simulation application for the LHCb experiment, eConf C0303241 (2003) TUMT003, arXiv:physics/0306035
41. D.J. Lange, The EvtGen particle decay simulation package. Nucl. Instrum. Meth. **A462**, 152 (2001)
42. GEANT4 collaboration, S. Agostinelli et al., GEANT4: A simulation toolkit, Nucl. Instrum. Meth. **A506**, 250 (2003)
43. The LHCb Collaboration, The Boole Project, http://lhcb-release-area.web.cern.ch/LHCb-release-area/DOC/boole/
44. The LHCb Collaboration, The Moore Project, http://lhcb-release-area.web.cern.ch/LHCb-release-area/DOC/moore/
45. The LHCb Collaboration, The Brunel Project, http://lhcb-release-area.web.cern.ch/LHCb-release-area/DOC/brunel/
46. The LHCb Collaboration, The DaVinci Project, http://lhcb-release-area.web.cern.ch/LHCb-release-area/DOC/davinci/

Chapter 3
Reconstruction and Selection

This chapter presents the criteria employed to create and select the signal candidates used in the fit to the CP-related observables. Candidates are selected exploiting the topology of the decays of interest, which have two well-separated secondary vertices. As LHCb suffers high background from low-q^2 QCD processes, the choice of a high performance selection is crucial to ensure an adequate level of background for the expected number of signal events. For this reason, a Multi Variate Analysis (MVA) is preferred to a cut-based analysis as it provides a higher signal efficiency for the same background rejection. Sections 3.1 and 3.2 present an overview on the decay modes considered and Sect. 3.3 describes the data and Monte Carlo (MC) simulated samples used in the analysis. From Sects. 3.4 to 3.8 the full selection chain is described, i.e. the preliminary reconstruction stage, the MVA classifier training, selection optimisation and evaluation of the expected total efficiencies. The chapter is then concluded by a study on the signal lineshapes in Sect. 3.9, the momentum rescaling in Sect. 3.10 and the particle identification (PID) calibration in Sect. 3.11. The analysis has been blinded for the suppressed $B \to DK$ mode only. All the other reconstructed modes, especially the most abundant favoured $B \to D\pi$ mode, provide useful information and internal cross-checks for the analysis itself.

3.1 Overview of the $B \to [hh]_D h$ Decay Modes

The analysis reconstructs the decays $B^\pm \to Dh^\pm$ with $D \to h^+h^-$, where h represents either a pion or a kaon (see Fig. 3.1). Note that no experimental distinction is possible in this case between the two *pure* intermediate flavour eigenstates D^0 and $\bar{D}^0$, since only the three-body $B \to [hh]_D h$ final state is accessible experimentally. The decay always proceeds through two interfering amplitudes, though this interference can be neglected for the most abundant combinations.

All sixteen $B^\pm \to [hh]_D h^\pm$ possible mass combinations are reconstructed and used in a simultaneous fit (see Chap. 5) to extract the CP observables defined in Sect. 1.6.3. Four neutral $[h^+h^-]_D$ combinations are possible: $[K^+K^-]_D$ and

P. Gandini, *Observation of CP Violation in $B^\pm \to DK^\pm$ Decays*,
Springer Theses, DOI: 10.1007/978-3-319-01029-8_3,

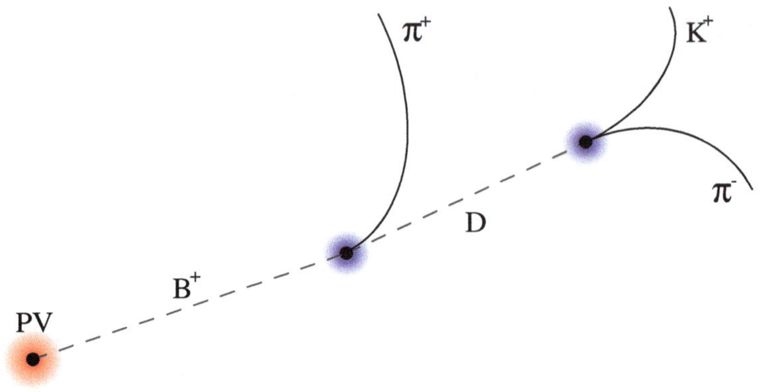

Fig. 3.1 Topology of signal $B \to [hh]_D h$ decays

$[\pi^+\pi^-]_D$ are *CP* eigenstates, while $[K^\pm\pi^\mp]_D$ and $[\pi^\pm K^\mp]_D$ notations label the favoured and the suppressed modes respectively. D candidates are then combined with another charged track in the event to form B candidates. In the rest of this document, the h that is emitted directly by the $B^\pm$ candidate will be denoted as the *bachelor* track, while the particles emitted by the intermediate D candidate will be denoted as the *daughter* tracks. The following notations are introduced for convenience:

- $B \to [KK]_D\pi$ and $B \to [KK]_D K$ for the $B^\pm \to [K^+K^-]_D h^\pm$ states.
- $B \to [\pi\pi]_D\pi$ and $B \to [\pi\pi]_D K$ for the $B^\pm \to [\pi^+\pi^-]_D h^\pm$ states.
- $B \to [K\pi]_D\pi$ and $B \to [K\pi]_D K$ for the favoured $B^\pm \to [K^\pm\pi^\mp]_D h^\pm$ states.
- $B \to [\pi K]_D\pi$ and $B \to [\pi K]_D K$ for the suppressed $B^\pm \to [\pi^\pm K^\mp]_D h^\pm$ states.

Of fundamental importance to the selection is the observation that the favoured and the suppressed modes only differ in the relative charge of the bachelor track and the daughter kaon, same sign being the favoured mode and opposite sign the suppressed one. Therefore the favoured mode can be used to model the suppressed signal mode and understand the background where possible. Same selection criteria are applied to all the $B \to [hh]_D\pi$ and $B \to [hh]_D K$ signal modes. In this way, most systematic uncertainties cancel out and only modest relative corrections on yields are needed (described in Sect. 5.3.1).

3.2 The First Favoured $B \to [K\pi]_D\pi$ Candidate

The very first favoured $B \to [K\pi]_D\pi$ candidate found by the LHCb detector in 2010 data (Event Number: 53382368, Run Number: 69648) is sketched in Fig. 3.1. The event is plotted in the transverse x-y plane (Table 3.1).

Table 3.1 Representation of the very first favoured $B \to [K\pi]_D\pi$ candidate in the transverse x–y plane

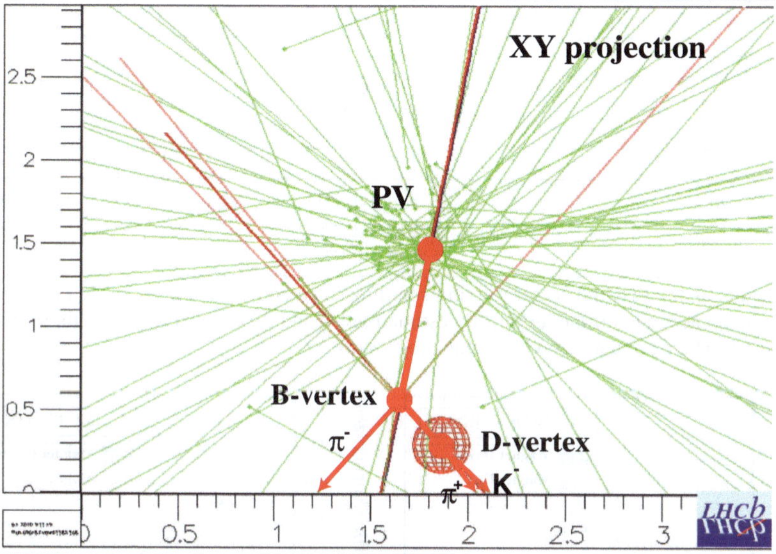

Particle	Mass (MeV/c^2)	p_T (MeV/c)	p (MeV/c)	z-vertex (mm)	y-vertex (mm)	x-vertex (mm)
B^-	5288.0	5289.4	184747	37.2446	−0.198245	−1.04383
D^0	1867.0	2648.3	107548	53.2506	−0.43799	−1.32324
Bachelor π^-	139.57	3980.15	77288.8	–	–	–
Daughter K^-	493.68	1182.96	22820.0	–	–	–
Daughter π^+	139.57	1465.93	84764.9	–	–	–
Primary Vtx	–	–	–	5.5857	−0.284	−0.146

The signal (rest of the event) tracks are plotted in red (green). The details of the reconstructed candidate are shown in the table

3.3 Datasets and MC Samples

The analysis uses the full 2011 LHCb dataset of $1.0\,\text{fb}^{-1}$, with almost an equal amount of integrated luminosity recorded in both magnet polarities, see Table 3.2. A sample of $35.7\,\text{pb}^{-1}$, collected in 2010 is also used in the analysis as an independent background sample for the MVA classifier training, as described in Sect. 3.5.1.

Large samples of Monte Carlo (MC) simulated events have been generated for all the signal decays and several background modes. Approximately one million simulated events for each $B \to [hh]_D h$ mode are used to study any possible difference in both selection efficiency and mass lineshape due to the different number of kaons and pions in the final state. MC simulated events are produced within the LHCb computing framework described briefly in Sect. 2.2.11. pp collisions are simulated with a tuned version of Pythia [2] requiring an average number of pp interactions

Table 3.2 Dataset collected by the LHCb detector and used in the analysis, split by magnet polarity. Integrated luminosities are expressed in pb^{-1}

2010	$\int \mathcal{L}\,\mathrm{d}t$	(rel. %)
Magnet down	18.2 ± 1.8	(51)
Magnet up	17.5 ± 1.8	(49)
Total	35.7 ± 3.6	(100)

2011	$\int \mathcal{L}\,\mathrm{d}t$	(rel. %)
Magnet down	589.3 ± 20.6	(58)
Magnet up	435.2 ± 15.2	(42)
Total	1024.5 ± 35.9	(100)

More details on the LHCb detector running conditions can be found in Sect. 2.2.1. These numbers come from an accurate offline luminosity estimation [1] and they thus differ from those presented in Table 2.1, which are derived with online procedures

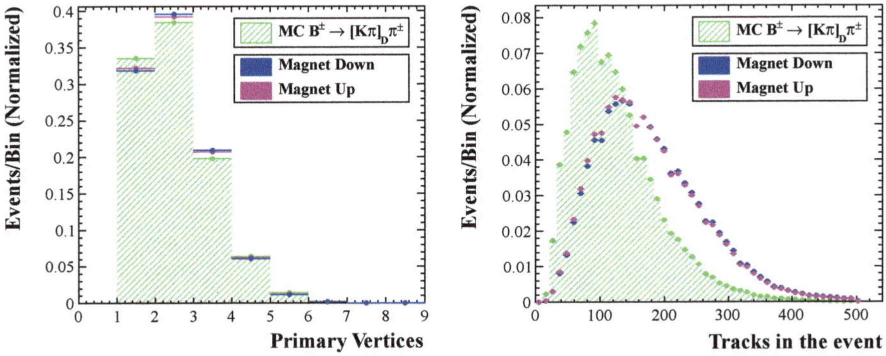

Fig. 3.2 Event variables. *Left* number of reconstructed Primary Vertices (PV) in the event. *Right* number of charged tracks reconstructed in the event. Both magnet *Down* and *Up* data samples have been collected with different running conditions during the year and they present a larger average number of tracks and PVs in the event compared to the simulated $B \to [K\pi]_D\pi$ events

per bunch crossing $\nu = 2$. This value is chosen to reproduce with reasonable approximation the 2011 LHCb running conditions, summarized in Sect. 2.2.1. Two event variables are plotted in Fig. 3.2 for data and MC simulated $B \to [K\pi]_D\pi$ events. The observed MC/data discrepancy depends both on the simulated detector running conditions and the description of the hadron fragmentation in the event and does not effect the *CP* observables, which are robust to this kind of event global variables.

EvtGen [3] is then used to model the particle decay chains and Geant4 [4] describes the interactions of the simulated particles in the detector. All MC generated signal tracks are required to lie in the LHCb detector acceptance (see Sect. 2.2). The geometrical acceptance $\varepsilon_{\mathrm{eff}}^{\mathrm{Geom}}$ is thus defined as:

$$\varepsilon_{\mathrm{eff}}^{\mathrm{Geom}} = \frac{n_{\mathrm{acc}}}{n_{\mathrm{gen}}}$$

Table 3.3 List of decay modes generated within the LHCb central MC production

Mode	n_{acc}	ε_{eff}^{Geom}
$B \to [K\pi]_D\pi$	1518243	0.1604 ± 0.0002
$B \to [K\pi]_D K$	1556991	0.1639 ± 0.0002
$B \to [KK]_D\pi$	1034996	0.1676 ± 0.0002
$B \to [KK]_D K$	1007996	0.1714 ± 0.0002
$B \to [\pi\pi]_D\pi$	1027494	0.1540 ± 0.0002
$B \to [\pi\pi]_D K$	1009996	0.1572 ± 0.0002
$B_u \to DX$ inclusive	40145017	0.3277 ± 0.0027
$B_d \to DX$ inclusive	40059091	0.3265 ± 0.0027
$B_s \to DX$ inclusive	10042907	0.3235 ± 0.0027
$\Lambda_b \to DX$ inclusive	10055701	0.3328 ± 0.0027
$B^+ \to J/\psi[\mu^+\mu^-]K^+$	4021979	0.1546 ± 0.0003
$B^+ \to [K\pi]_D\mu^+\nu$	158896	0.1668 ± 0.0005
$\Lambda_b \to [Kp\pi]_{\Lambda_c}h$	2033496	0.1634 ± 0.0003
$B_s \to [K\pi]_{D^0}K^*$	505498	0.1569 ± 0.0004
$B^+ \to \pi^+\pi^-K^+$	1690094	0.1606 ± 0.0005
$B^+ \to K^+K^-K^+$	1660492	0.1698 ± 0.0005
$B^+ \to \pi^+\pi^-\pi^+$	2220989	0.1557 ± 0.0005
$B^+ \to K^+K^-\pi^+$	157100	0.1649 ± 0.0005

The first part of the table reports the six different $B \to [hh]_D h$ signal modes, while the second part the inclusive and exclusive background modes used in Chap. 4. For the $B \to DX$ inclusive samples, signal-related tracks are only required to be forward in the detector direction, thus the higher acceptance efficiency

where n_{acc} is the number of geometrically-accepted events with all three signal charged tracks inside the detector acceptance and n_{gen} the total number of generated events in the 4π solid angle. A summary of all the simulated signal and background modes used in the analysis is presented in Table 3.3.

3.4 Stripping

After the track reconstruction (see Sect. 2.2.9), B candidates are built and filtered according to specific loose requirements. The purpose of this preliminary selection, usually referred to as *stripping*, is to reduce the computing resources, both CPU time and storage space, needed for the end-user analyses and thus provide the analysts with manageable data samples. Candidates are taken from the `StrippingB2D0PiD2HH` and `StrippingB2D0KD2HH` lines which are included in the 2011 `Beauty2Charm` stripping stream. The `Beauty2Charm` stream is large inclusive collection of lines for $B \to DX$ hadronic decays, including more than 800 different final states. Its basic strategy is to use the events accepted by the hadronic topological triggers as a reasonable starting point, then do most background rejection by means of loose cuts on displaced B and D

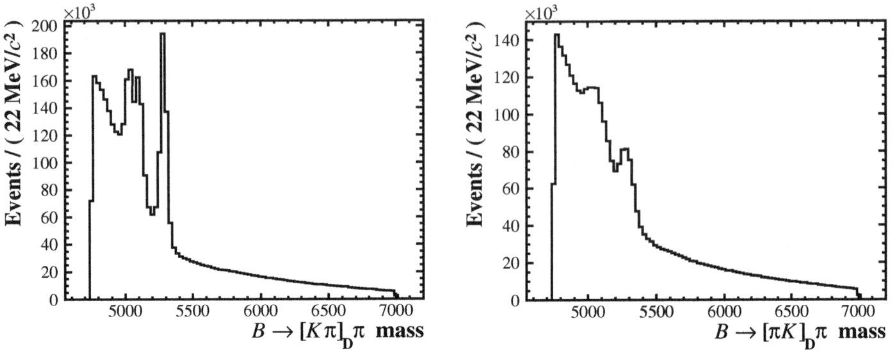

Fig. 3.3 B mass distributions of signal candidates from the stripping on data. *Left* favoured $B \rightarrow [K\pi]_D\pi$ candidates. *Right* suppressed $B \rightarrow [\pi K]_D\pi$ candidates. The signal-like peak in the *right plot* is due to favoured → suppressed-mode cross feed (Sect. 4.5.3)

candidates, e.g. requiring good quality vertices ($\chi^2_{\text{vtx}}(B, D) < 10$) and a minimum B proper lifetime of 0.2 ps. $D \rightarrow hh$ combinations are required to lie within $[-100, +100]$ MeV/c^2 from the nominal D^0 mass [5] while $B \rightarrow DX$ combinations must have $4750 < m_B < 7000$ MeV/c^2. Events are rejected before any computation if the number of tracks in the event is greater than 500 (too many track combinations possible). All signal tracks must have a good reconstruction $\chi^2_{\text{track}} < 4$ (see Sect. 2.2.9), in order to reduce CPU usage on fake track combinations. Candidates are saved in the stream if they pass a loose cut on a simple Boosted Decision Tree (BDT) discriminator that uses the momentum of the B candidate, its flight distance significance and the sum of all displaced χ^2_{vtx} in the decay chain as input variables. The relatively simple architecture of this BDT is chosen to ensure a fast response and keep under control the total computing time per single event. Selected candidates are saved into a separate dataset and used as input to a more elaborate MVA selection developed specifically for the $B \rightarrow [hh]_Dh$ final state in Sect. 3.5. The output of the stripping is plotted in Fig. 3.3 for the favoured and suppressed B candidates. Typical reconstruction and stripping efficiencies for the $B \rightarrow [hh]_Dh$ modes are of 5.6 % (see Sect. 3.8).

3.5 Selection: A MVA Approach

This section describes a multivariate (MVA) selection for signal $B \rightarrow [hh]_Dh$ events. This selection exploits the particular decay topology of these modes and the same classifier is used for all the $B \rightarrow [hh]_Dh$ modes, although the optimal working point can differ for the *CP* modes. A MVA approach is preferred to a cut-based selection, as used in the early stages of the analysis [6], in order to ensure an excellent rejection of combinatorial background while providing reasonable signal efficiencies. Furthermore, MVA methods can exploit correlations between similar types of input variables and maximise signal efficiency. Several different MVA classifiers have been investigated, including Boosted Decision Trees (BDT) and Neural Networks (NN), within the TMVA package [7]. The BDT classifier has been preferred for its fast response and its reliable property of *automatic ranking* of the input variables in the boosting process, which makes BDT a robust classifier with respect to overtraining. Furthermore, the hadronic HLT2 trigger (see Sect. 2.2.8) is based on a BDT classifier decision. Several configurations of the input variables of the BDT, the training samples and the classifier parameters (e.g number of trees in the forest, number of nodes in each tree, learning curve, etc.) have been tested. This section documents only the best configuration found and therefore that used for the analysis. Before proceeding to the training and testing of the MVA classifier, sanity cuts have been introduced on all the used samples (both MC and data): the stripping requires D daughter tracks to have $p_T > 250\,\mathrm{MeV}/c$ but this requirement is tightened to $0.5 < p_T < 10\,\mathrm{GeV}/c$ and $5 < p < 100\,\mathrm{GeV}/c$ for the bachelor tracks. The high-momentum $p < 100\,\mathrm{GeV}/c$ cut rejects tracks with a non-satisfactory RICH detector performance (see Sect. 2.2.10), as for ultra-high momentum tracks the pion and kaon mass hypotheses are indistinguishable.

3.5.1 Input Samples and Variables

The BDT discriminator is trained using a large MC simulated sample of $B \to [K\pi]_D h$ signal events (truth matched), while background events from the D sidebands of the 2010 $35.7\,\mathrm{pb}^{-1}$ sample (defined as $35 < |m_D - \mu_D| < 100\,\mathrm{MeV}/c^2$) provide an independent sample of combinatorial background. A small contamination from misidentified $[KK]_D$ and $[\pi\pi]_D$ events is expected in the sideband sample as described in Sect. 4.5.1, although negligible compared to the size of the combinatorial background. This choice of the background sample has the additional benefit that the optimisation procedure is fully independent of the 2011 $1.0\,\mathrm{fb}^{-1}$ dataset used for the analysis. The BDT discriminator uses the following properties of both B and D candidates and tracks, the logarithm of the variable being preferred when the variable distribution presents long tails or it is distributed on a wide range:

- The transverse momenta: $p_\mathrm{T}(B)$, $p_\mathrm{T}(D)$, $p_\mathrm{T}(bach)$ and two $p_\mathrm{T}(D\ dau)$.
- The minimum impact parameters: $\chi^2_{IP}(B)$, $\chi^2_{IP}(D)$ and two $\chi^2_{IP}(D\ dau)$. Impact parameters are defined in Fig. 3.4.
- The maximum distance of closest approach (DOCA) of either the D meson or bachelor track with respect to the B vertex and the maximum DOCA of either daughters with respect to the D vertex.
- The proper decay times τ_B and τ_D of the candidates and their separation significance $S_{FD}(B)$ and $S_{FD}(D)$ from the PV.
- The $L_{xy}(B)$ and $L_{xy}(D)$ flight distances in the transverse plane.
- The vertex quality of both B and D candidates: $\chi^2_{vtx}(B)$ and $\chi^2_{vtx}(D)$

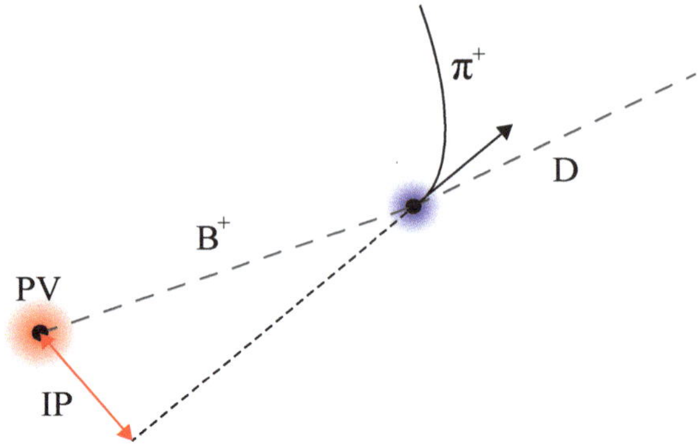

Fig. 3.4 Definition of the Impact Parameter variable (IP)

- The pointing angle $\cos(\theta_B)$, which is the angle between the B-candidate's momentum vector and a line connecting the PV to its decay vertex.

Information coming from the rest of the event, e.g. the tracks in the event that do not belong to the $B \to [hh]_D h$ candidate, is also employed in the training through the asymmetry variable A_{p_T} described in the next paragraph. PID information is not used in the training in order not to introduce dependencies on any particular $[hh]_D$ mode. An overview of the distributions of the MVA input variables for the MC simulated signal and sideband background samples is given in Appendix A , while Table 3.4 reports the ranking of the input variables used in the BDT training. It is interesting to note that the top three variables are all related to the impact parameters of the reconstructed particles.

A_{p_T} Isolation Variable

Information from the rest of the event is employed via an isolation variable that considers the imbalance of p_T in the rest of the event. The variable is constructed building a cone, defined in the $\phi - \eta$ coordinates, with ϕ measured in radians, and counting the charged tracks lying inside this cone that do not contribute to the signal candidate. The imbalance is then defined as:

$$A_{p_T} = \frac{p_T(B) - [\sum_n p]_T}{p_T(B) + [\sum_n p]_T},$$

where $[\sum_n p]_T$ is the sum over the n tracks within the cone, excluding the three signal tracks. This variable, widely used in hadronic colliders [8], provides additional information virtually independent from the signal topology, as the rest of the event is mostly used to build the variable. Since MC simulated events are used in the training, a good MC/data agreement is required in the description of the full event inside the LHCb detector. This study is presented in Appendix B for different aperture angles θ_{cone}; the A_{p_T} distributions appear well described in the simulation. The A_{p_T}

Table 3.4 Ranking of the BDT input variables after the training process

Rank	Variable	Variable importance
1	$\log[\chi^2_{IP}(B)]$	1.042×10^{-1}
2	$\min[\log[\chi^2_{IP}(D\ dau)]]$	9.248×10^{-2}
3	$\log[\chi^2_{IP}(D)]$	8.437×10^{-2}
4	$p_T(B)$	7.706×10^{-2}
5	A_{p_T}	7.369×10^{-2}
6	$p_T(bach)$	7.265×10^{-2}
7	$\chi^2_{vtx}(B)$	6.507×10^{-2}
8	$\min[p_T(D\ dau)]$	6.064×10^{-2}
9	$p_T(D)$	5.633×10^{-2}
10	$\chi^2_{vtx}(D)$	5.559×10^{-2}
11	τ_D	5.217×10^{-2}
12	$\log[S_{FD}(B)]$	5.087×10^{-2}
13	$\log[\cos(\theta_B)]$	3.999×10^{-2}
14	$\log[S_{FD}(D)]$	3.894×10^{-2}
15	$\mathrm{DOCA}(B)$	3.161×10^{-2}
16	$L_{xy}(B)$	1.798×10^{-2}
17	$\mathrm{DOCA}(D)$	1.645×10^{-2}
18	τ_B	8.068×10^{-3}
19	$L_{xy}(D)$	1.834×10^{-3}

The ranking is a measure of the importance of each variable in the BDT discriminator and it is calculated by counting how often the variables are used in the decision process and weighting their specific separation power as described in [9]. Note that weights sum to 1

variable is also plotted in Fig. 3.5 for different aperture angles with the chosen value of $\theta_{\text{cone}} = 1.5$ giving the best separation power between the signal and background samples.

3.5.2 Performance and Optimal Cut Evaluation

The BDT classifier distribution is shown in Fig. 3.6, with its calculated Relative Operating Characteristic (ROC) curve. A data-driven approach is chosen to identify the optimal working point on the ROC curve, e.g the best cut on the BDT variable in order to extract the suppressed $B \to [\pi K]_D K$ signal mode. The most abundant favoured $B \to [K\pi]_D\pi$ data sample in thus exploited to estimate signal and background rates, once the signal peak is properly scaled. For this purpose, several fits are performed on the $B \to [K\pi]_D\pi$ sample (after preliminary selection only) at different BDT cuts using a simple model including a modified Gaussian (Sect. 3.9) for the signal component, a linear polynomial for the combinatorial background and an effective PDF to account for the partially reconstructed backgrounds below the B mass peak (introduced in Sect. 4.3). An example plot of this fit is shown in Fig. 3.7.

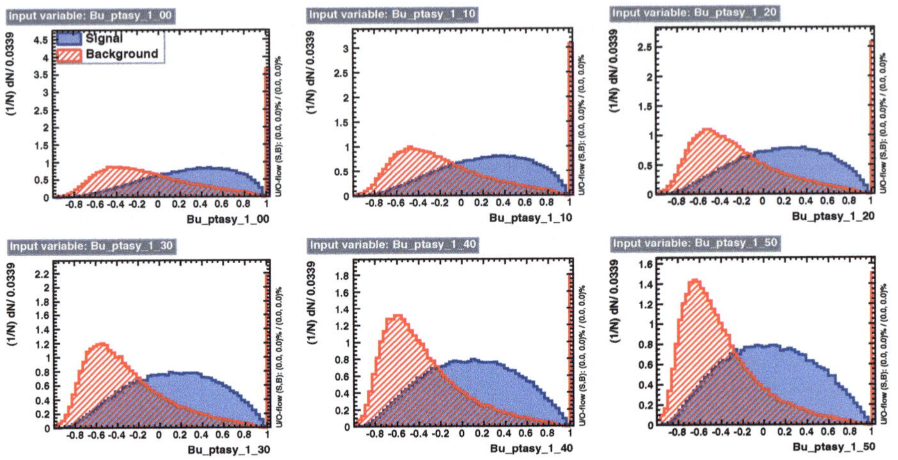

Fig. 3.5 A_{p_T} variable distribution, at different cone apertures, for the signal and background samples used in the BDT discriminator training. *Blue* signal simulation. *Red* data 2010 sidebands. The x-axis labels can be decoded as follows: `Bu_ptasy` is the A_{p_T} variable followed by the opening angle value expressed in radians (only apertures in the range 1.0−1.50 radians are plotted)

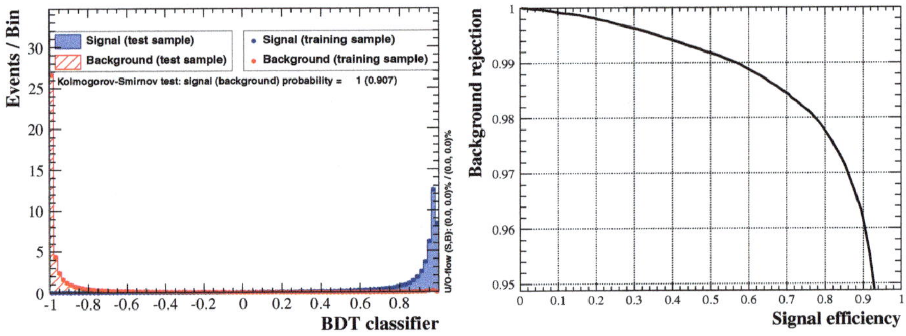

Fig. 3.6 *Left* classifier distribution for signal and background events. The BDT response peaks towards −1 for background and +1 for signal candidates. *Right* ROC curve (signal efficiency vs. background rejection) for background rejection rates >0.95

The number of signal (n_S) and background (s_B) events for every given BDT cut is shown in Fig. 3.8.

The optimal cut value is chosen assuming that the level of combinatorial background is the same in the favoured and the suppressed mode and scaling the signal peak size according to the anticipated suppressed branching fraction (see Sect. 1.6.3). The significance σ is calculated as

$$\sigma = \frac{n_S}{\sqrt{n_S + n_B}} \quad \text{with} \quad n_S = \frac{\mathcal{B}_{\text{target}}}{\mathcal{B}_{\mathrm{B}\to[\mathrm{K}\pi]_{\mathrm{D}}\pi}}\, n_{\mathrm{B}\to[\mathrm{K}\pi]_{\mathrm{D}}\pi}$$

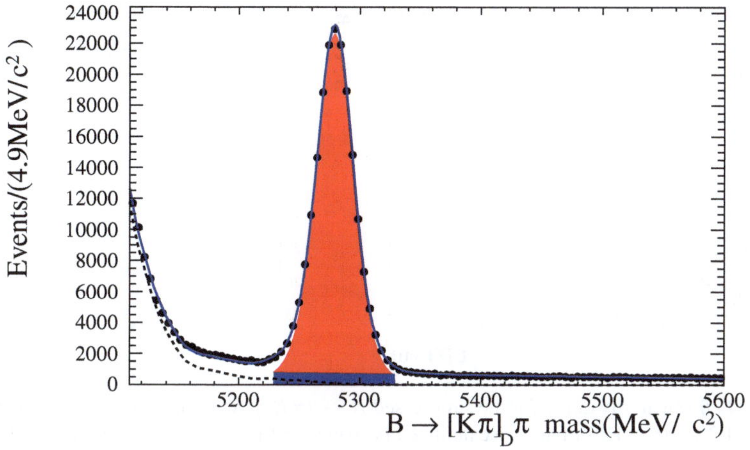

Fig. 3.7 Schematic view of the simple fit on the $B \to [K\pi]_D\pi$ mode used to extract n_S and n_B from data. The *dotted line* represents partially reconstructed backgrounds (see Sect. 4.3.2)

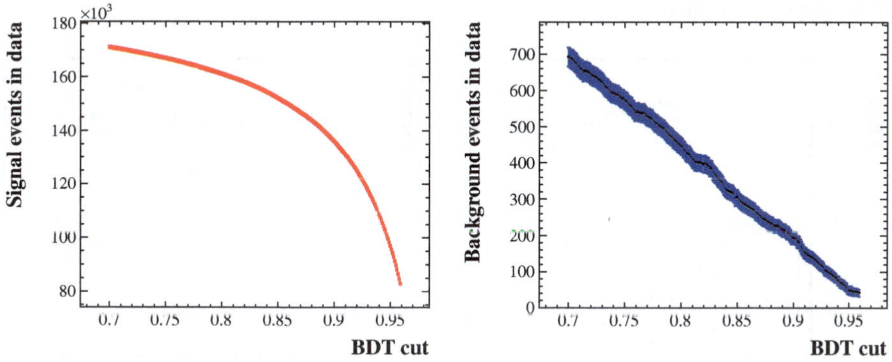

Fig. 3.8 *Left* number of $B \to [K\pi]_D\pi$ signal events fitted on data for a given BDT cut. *Right* number of combinatorial background events

where *target* is the desired $B \to [hh]_D h$ decay mode (about 2×10^{-7} for the suppressed $B \to [\pi K]_D K$ mode) and $\mathcal{B}_{B\to[K\pi]_D\pi} = 1.87 \times 10^{-4}$ [5].

The significance serves as a metric for the optimisation as shown in Fig. 3.9. The optimal cut is found to be BDT > 0.92 for the suppressed $B \to [\pi K]_D K$ mode and this is also applied to the favoured $B \to [K\pi]_D h$ decay. The *CP* modes are more abundant than the suppressed mode and have less backgrounds, so a looser cut of BDT > 0.80 is applied.

Several tests have been performed in order to check the robustness of the BDT selection:

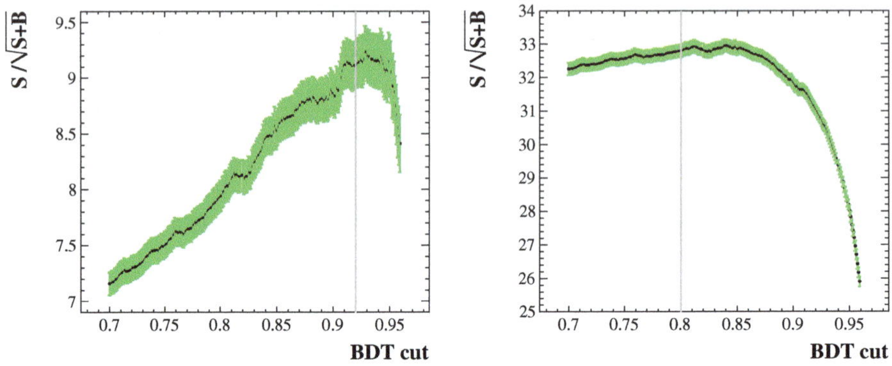

Fig. 3.9 *Left* sensitivity curve for the suppressed $B \to [\pi K]_D K$ mode assuming a total branching fraction of 2×10^{-7}. Error bars are calculated propagating the uncertainties on the fitted values of $n_{B\to[K\pi]_D\pi}$ and n_B, the combinatorial background yield. The optimal cut for the selection is chosen to be at the lower edge of the wide plateau in the region of BDT > 0.90 and it is indicated with a vertical line. *Right* sensitivity curve imposing a signal branching fraction of about 1.5×10^{-6} for the $B \to [KK]_D K$ decay. The maximum of the curve is shifted to lower values, allowing a looser BDT cut for the *CP* modes

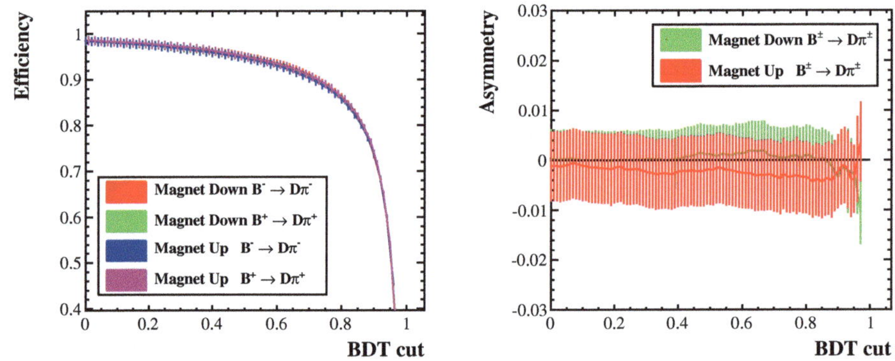

Fig. 3.10 *Left* Efficiencies of the BDT cuts for $B \to [K\pi]_D\pi$ events as fitted on data. The sample is split by charge and magnet polarity (four subsamples in total). *Right* charge asymmetry split by magnet polarity. As expected, the induced asymmetry is compatible with zero

- Since no charge dependent variable is introduced in the training and both $B^{\pm}$ candidates are used, no apparent *CP* asymmetry is expected to be induced by the selection. This is double-checked on the full BDT range in Fig. 3.10.
- The shape parameters of the signal peak are plotted as function of the MVA classifier. No significant distortion is introduced in peak over the full BDT range as illustrated in Fig. 3.11. The width and left tail parameters of the peak are found to vary, but variations are within one sigma. Those parameters are left floating in the fitter and are the same for the favoured and suppressed modes.

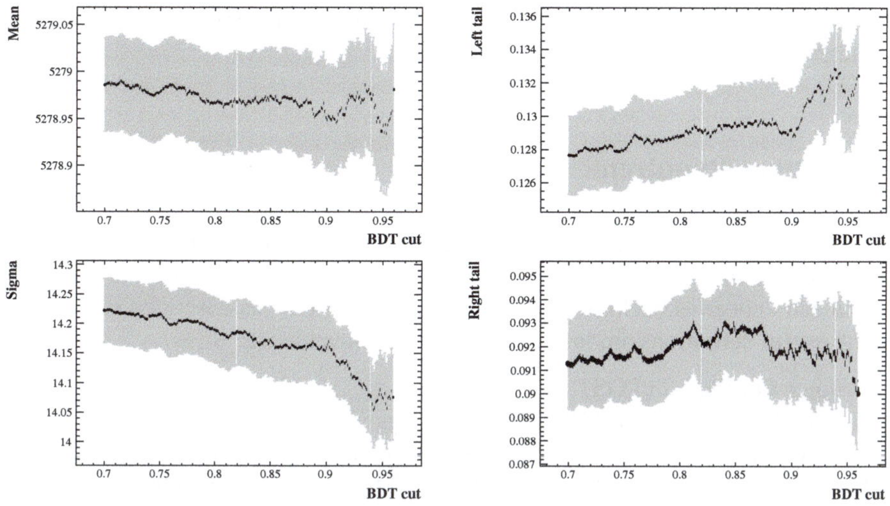

Fig. 3.11 Signal PDF parameters with respect to BDT cut. The signal component is parameterized with a modified Gaussian, as described in Sect. 3.9, which depends on four parameters: mean, sigma and the two tails. The decrease in the fitted width with tightening BDT cut is consistent with a selection that favours the most accurate decay reconstruction. The left tail parameter rises to compensate for the genuine radiative tail. However, all variations are within one sigma deviation and so are not cause of concern

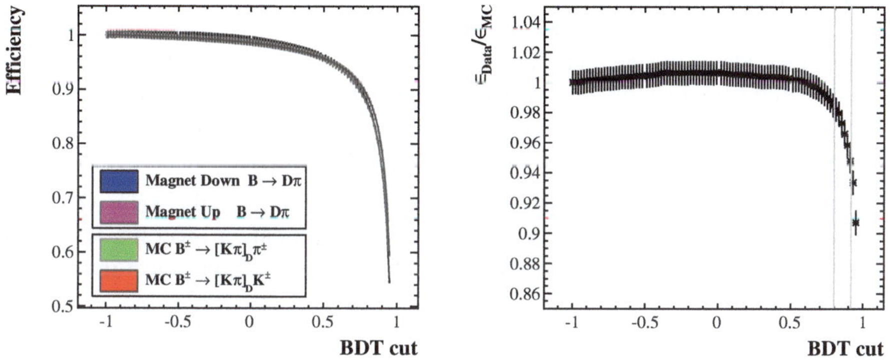

Fig. 3.12 *Left* efficiencies for MC and data samples. *Right* correction factor $\epsilon_{data}/\epsilon_{MC}$. Efficiencies are normalized to 1 for BDT $= -1$, e.g no MVA cut applied

- The BDT performance is evaluated both on data and MC simulated events and MC/data efficiencies are presented in Fig. 3.12. BDT is found to perform better on simulated events than on data in the range of interest. This is expected since $B \rightarrow [K\pi]_D h$ MC events are used in the training procedure, while data is only used in the background description and the optimal cut evaluation. These MC/data differences are used to correct absolute estimations based on MC simulated events.

However, it should be noted that this sort of corrections cancels out almost completely when providing relative estimations.

3.6 Trigger Requirements

Specific trigger requirements on the selected candidates are applied in both reconstruction and offline selection. This is mainly useful for removing muon triggers which can appear as backgrounds peaking at the B mass, as described in Sect. 4.2. Some useful definitions are introduced here for convenience:

- **Trigger On Signal (TOS)**
  B candidates are defined TOS if they have triggered a particular trigger line, e.g. the `L0Hadron_TOS` requirement is satisfied only if the reconstructed signal particle has fired the L0 hadron trigger line (Sect. 2.2.8). This means that the signal tracks alone are sufficient to select the event via the particular line in question.
- **Trigger Independent of Signal (TIS)**
  A candidate is defined TIS if the event has been triggered by a particular line but independently of the candidate itself. This happens when no overlap is found between the signal candidate tracks and the object that has triggered the line under consideration. TIS events provide an unbiased sample of triggered events since the B candidate has played no part in the trigger decision.
- **Trigger On Both (TOB)**
  TOB candidates are neither TOS nor TIS objects. This can happen when only part of the signal candidate is used to trigger the given line. Events of this type are difficult to describe experimentally, as they usually depend on fake tracks reconstructed at the trigger stage. TOB events are removed from the analysis.

It should be noted that TOS and TIS categories are not mutually exclusive since a candidate can be both TIS and TOS, i.e it has fired a particular trigger line which has been fired at the same time by a completely independent trigger object in the same event.

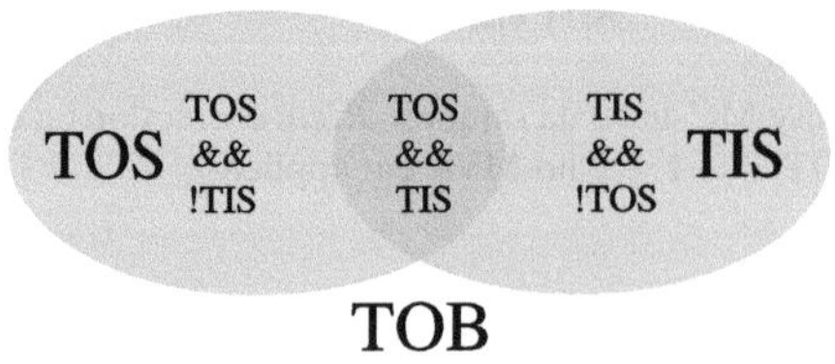

Candidates are accepted only if they have (`L0Hadron_TOS` *or* `L0Global_TIS`) at the L0 level, with `Global` being the inclusive merge of all the physics trigger lines in *or* mode. This use of trigger-selected events removes most of charmonium events like $B^{\pm} \rightarrow J/\psi K^{\pm}$ decays (see Sect. 4.2.1) which are triggered by the `L0Muon_TOS` lines. Candidates are then selected requiring `Hlt1Track_TOS` and

Hlt2Topo_TOS for the two, three and four-body HLT2 topological lines in *or* configuration. Efficiencies of trigger requirements have been studied on simulated signal events reproducing the 2011 running conditions of the experiment, described in Sect. 2.2.1. The 0x760037 TCK has been chosen for all the simulated events: this Trigger Configuration Key has been used to collect about 30 % of the total dataset used in this analysis and it is one of the most representative of the 2011 data taking period. Trigger efficiencies for all the $B \to [hh]_D h$ modes, defined as $n_{\text{triggered}}/n_{\text{offline}}$, are shown in Table 3.5.

The choice of selecting L0Global_TIS, as well as L0Hadron_TOS candidates, increases the signal yield by 20 %. This trigger strategy is common for all $B \to DX$ hadronic analyses in the LHCb collaboration. The trigger requirement efficiency is checked in data with a procedure similar to the one presented in Sect. 3.5.2. The $B \to [K\pi]_D\pi$ favoured control sample, once the full selection is applied, is fitted with and without TIS-TOS requirements and the obtained relative efficiencies are shown in Table 3.6. It should be noted that this correction cancels in the *CP* physical observables (which are ratios of partial widths) and only small relative corrections are introduced (Sect. 5.3.1). Data results on the favoured $B \to [K\pi]_D\pi$ mode are fully compatible with the results obtained in MC events.

3.7 Multiple Candidate Arbitration

Multiple candidate arbitration is performed after all other selection criteria are applied. This ensures that no candidate appears twice, anywhere in the signal extraction fit. The nominal D mass window cut and $\Delta LL^{dau}_{K/\pi}$ PID requirements on the daughter tracks remove the possibility of one candidate to appear in two different $[hh]_D$ modes. In practice, multiple candidates are rare, since clone tracks (tracks with multiple detector hits in common) are removed during the LHCb track reconstruction, and arbitration is only needed 19 times in a fitted sample which contains about 133 thousand events. In this cases, the B candidate with a vertex χ^2/n_{dof} closest to 1 is chosen and in case of equal χ^2/n_{dof} the one with m_D closest to the nominal D^0 mass value is preferred.

3.8 Total Efficiencies and Expected Yields

The total selection efficiencies are summarized in Table 3.7 for all the $B \to [hh]_D h$ signal modes. Each cut is reported with its absolute efficiency and relative to all the previous cuts. For completeness, total efficiencies also include the effects of all the cuts introduced in Chap. 4 in order to suppress peaking backgrounds. MC efficiencies are used to estimate the expected number of signal events for each $B \to [hh]_D h$ mode. Expected yields can be obtained as follows:

Table 3.5 Efficiencies of TIS-TOS trigger requirements for $B \to [hh]_D h$ MC simulated and truth matched events

	$B \to [K\pi]_D \pi$		$B \to [K\pi]_D K$	
	MC evts	*Efficiency* $\varepsilon_{\rm eff}^{\rm TISTOS}$ (%)	MC evts	*Efficiency* $\varepsilon_{\rm eff}^{\rm TISTOS}$ (%)
Selection + Bkg vetoes	39829	–	40041	–
`L0Hadron_TOS`	30993	(77.82 ± 0.59)	31381	(78.37 ± 0.59)
`L0Global_TIS`	15684	(39.38 ± 0.37)	15819	(39.51 ± 0.37)
`L0Hadron_TOS` *and* `L0Global_TIS`	8146	(20.45 ± 0.25)	8394	(20.96 ± 0.25)
`L0Hadron_TOS` *or* `L0Global_TIS`	38531	(96.74 ± 0.69)	38806	(96.92 ± 0.69)
`Hlt1Track_TOS`	38464	(96.57 ± 0.69)	38738	(96.75 ± 0.69)
`Hlt2Topo2_TOS`	34807	(87.39 ± 0.64)	35290	(88.13 ± 0.64)
`Hlt2Topo3_TOS`	28075	(70.49 ± 0.55)	28347	(70.79 ± 0.55)
`Hlt2Topo4_TOS`	46	(0.12 ± 0.02)	54	(0.13 ± 0.02)
Total (all requirements)	37943	(95.26 ± 0.68)	38275	(95.59 ± 0.68)
	$B \to [KK]_D \pi$		$B \to [KK]_D K$	
	MC evts	*Efficiency* $\varepsilon_{\rm eff}^{\rm TISTOS}$ (%)	MC evts	*Efficiency* $\varepsilon_{\rm eff}^{\rm TISTOS}$ (%)
Selection + Bkg vetoes	33487	–	32284	–
`L0Hadron_TOS`	25658	(76.62 ± 0.64)	24745	(76.65 ± 0.65)
`L0Global_TIS`	13787	(41.17 ± 0.42)	13461	(41.70 ± 0.43)
`L0Hadron_TOS` *and* `L0Global_TIS`	6980	(20.84 ± 0.27)	6927	(21.46 ± 0.28)
`L0Hadron_TOS` *or* `L0Global_TIS`	32465	(96.95 ± 0.76)	31279	(96.89 ± 0.77)
`Hlt1Track_TOS`	32399	(96.75 ± 0.75)	31219	(96.70 ± 0.77)
`Hlt2Topo2_TOS`	29111	(86.93 ± 0.70)	28100	(87.04 ± 0.71)
`Hlt2Topo3_TOS`	24909	(74.38 ± 0.62)	24095	(74.63 ± 0.64)
`Hlt2Topo4_TOS`	38	(0.11 ± 0.02)	56	(0.17 ± 0.02)
Total (all requirements)	32079	(95.80 ± 0.75)	30950	(95.87 ± 0.76)
	$B \to [\pi\pi]_D \pi$		$B \to [\pi\pi]_D K$	
	MC evts	*Efficiency* $\varepsilon_{\rm eff}^{\rm TISTOS}$ (%)	MC evts	*Efficiency* $\varepsilon_{\rm eff}^{\rm TISTOS}$ (%)
Selection + Bkg vetoes	34664	–	33979	–
`L0Hadron_TOS`	26296	(75.86 ± 0.62)	25850	(76.08 ± 0.63)
`L0Global_TIS`	14413	(41.58 ± 0.41)	14033	(41.30 ± 0.41)
`L0Hadron_TOS` *and* `L0Global_TIS`	7104	(20.49 ± 0.27)	6900	(20.31 ± 0.27)
`L0Hadron_TOS` *or* `L0Global_TIS`	33605	(96.94 ± 0.74)	32983	(97.07 ± 0.75)
`Hlt1Track_TOS`	33525	(96.71 ± 0.74)	32905	(96.84 ± 0.75)
`Hlt2Topo2_TOS`	29816	(86.01 ± 0.68)	29285	(86.19 ± 0.69)
`Hlt2Topo3_TOS`	23215	(66.97 ± 0.57)	22990	(67.66 ± 0.58)
`Hlt2Topo4_TOS`	34	(0.10 ± 0.02)	60	(0.18 ± 0.02)
Total (all requirements)	32911	(94.94 ± 0.73)	32345	(95.19 ± 0.74)

All efficiencies are relative to selected events after both MVA selection (Sect. 3.5) and background vetoes (Sect. 4.5) are applied to all candidates. Errors are obtained propagating the pure statistical $\sqrt{n}$ Poissonian error on the number of events. The last row reports the total TIS-TOS efficiency for the chosen requirement in the analysis (`L0Hadron_TOS` *or* `L0Global_TIS`) *and* (`Hlt2Topo_TOS` for the two, three and four-body topological lines in *or* configuration)

Table 3.6 Trigger efficiencies obtained fitting the $B \to [K\pi]_D\pi$ favoured control sample (with and without trigger requirements)

	Fitted events	*Efficiency* $\varepsilon_{\mathrm{eff}}^{\mathrm{TISTOS}}$ (%)
Selection + Bkg vetoes	71586 ± 282	–
L0Hadron_TOS	48115 ± 228	(67.21 ± 0.41)
L0Global_TIS	33646 ± 193	(47.00 ± 0.33)
L0Hadron_TOS *and* L0Global_TIS	13100 ± 120	(18.30 ± 0.18)
L0Hadron_TOS *or* L0Global_TIS	68652 ± 274	(95.90 ± 0.54)
Hlt1Track_TOS	68489 ± 273	(95.67 ± 0.54)
Hlt2Topo2_TOS	62143 ± 260	(86.81 ± 0.50)
Hlt2Topo3_TOS	50286 ± 233	(70.25 ± 0.43)
Hlt2Topo4_TOS	72 ± 9	(0.10 ± 0.01)
Total (all requirements)	67607 ± 271	(94.44 ± 0.53)

The total efficiency $(94.44 \pm 0.53)\%$ is compatible to those obtained in Table 3.5 for MC $B \to [K\pi]_D\pi$ events

$$n_{B\to[hh]_Dh} = \mathcal{L}_{int} \cdot \sigma_{(pp\to b\bar{b}X)} \cdot 2f_{\mathrm{had}} \cdot \mathcal{B}_{B\to[hh]_Dh} \cdot \varepsilon_{\mathrm{eff}}^{\mathrm{Geom}} \cdot \varepsilon_{\mathrm{eff}}^{\mathrm{Sel}} \cdot \varepsilon_{\mathrm{eff}}^{\mathrm{PID_{dau}}} \cdot f_{\mathrm{corr}}$$

where

- $\mathcal{L}_{int}$ is the integrated luminosity (Table 3.2): 1024 ± 36 (3.5 % rel.),
- $\sigma_{(pp\to b\bar{b}X)}$ is the production cross-section taken from [10]: $284 \pm 20 \pm 49$ mb (18.6 % rel.),
- f_{had} is the hadronisation factor taken from [11, 12]: 0.4,
- $\mathcal{B}_{B\to[hh]_Dh}$ is the measured branching fraction $\mathcal{B}_{B\to[hh]_Dh} = \mathcal{B}_{B\to Dh}\mathcal{B}_{D\to hh}$ from [5],
- $\varepsilon_{\mathrm{eff}}^{\mathrm{Geom}}$ is the LHCb geometrical acceptance for each mode (Table 3.3),
- $\varepsilon_{\mathrm{eff}}^{\mathrm{Sel}}$ is the total selection efficiency or each mode (Table 3.8),
- $\varepsilon_{\mathrm{eff}}^{\mathrm{PID_{dau}}}$ is the daughter $\Delta LL_{K/\pi}$ cut efficiencies as fitted on data,
- f_{corr} is a data/MC correction factor (Fig. 3.12).

Branching fraction calculations ignore *CP* violation and are the simple product of two branching fractions, $\mathcal{B}_{B\to Dh} \cdot \mathcal{B}_{D\to hh}$. The relative uncertainty on the expected yields is about 25 % and suffers mainly from the high systematic uncertainty on the production cross-section (19% rel.). The expected suppressed $B \to [\pi K]_Dh$ yields can be obtained from the favoured $B \to [K\pi]_Dh$ yields assuming $R_{\mathrm{ADS}(\pi)} \sim 0.003$ and $R_{\mathrm{ADS}(K)} \sim 0.014$, which are reasonable assumptions as taken from the HFAG group [13]. The $n_{B\to[hh]_Dh}$ yields are summarized in Table 3.7.

3.9 Resolution and Mass Lineshapes

This section gives an overview of the properties of the $B \to [hh]_Dh$ signal components: the D and B mass peak resolutions, their lineshapes to be used in the fitter and the definition of the signal mass windows.

3.9.1 Correctly Reconstructed $B \to [hh]_D h$ Peaks

Correctly reconstructed signal candidates are expected to peak at the nominal B and D masses. The dominant Gaussian-like signal peaks are modeled with a modified Gaussian formula that contains two tail terms introduced to describe possible non-Gaussian tails. This modified Gaussian shape is defined as follows:

Table 3.7 Tables of efficiencies for $B \to [hh]_D h$ MC simulated events

	$B \to [K\pi]_D\pi$			$B \to [K\pi]_D K$		
	MC evts	*Absolute eff.(%)*	*Relative eff.(%)*	MC evts	*Absolute eff.(%)*	*Relative eff.(%)*
MC simulated	1518243	100.0	–	1556991	100.0	–
Preliminary	85589	(5.64 ± 0.02)	(5.64 ± 0.02)	87962	(5.65 ± 0.02)	(5.65 ± 0.02)
p and p_T cuts	72801	(4.80 ± 0.02)	(85.06 ± 0.43)	74339	(4.77 ± 0.02)	(84.51 ± 0.42)
BDT > 0.92	53577	(3.53 ± 0.02)	(73.59 ± 0.42)	54439	(3.50 ± 0.02)	(73.23 ± 0.41)
Charmonium veto	53577	(3.53 ± 0.02)	(100.00 ± 0.61)	54437	(3.50 ± 0.02)	(100.00 ± 0.61)
[*dau, bach*] Cross feed	53172	(3.50 ± 0.02)	(99.24 ± 0.61)	54052	(3.47 ± 0.02)	(99.29 ± 0.60)
Fav $\to$ Sup cross feed	49153	(3.24 ± 0.01)	(92.44 ± 0.58)	49821	(3.20 ± 0.01)	(92.17 ± 0.57)
$FD_{[z_D^{vtx}-z_B^{vtx}]} > 2$	40411	(2.66 ± 0.01)	(82.21 ± 0.55)	40613	(2.61 ± 0.01)	(81.52 ± 0.54)
$\lvert m_D - \mu_D \rvert <$ 25 MeV/c^2	39829	(2.62 ± 0.01)	(98.56 ± 0.70)	40041	(2.57 ± 0.01)	(98.59 ± 0.69)
Trigger	37949	(2.50 ± 0.01)	(95.26 ± 0.68)	38275	(2.46 ± 0.01)	(95.59 ± 0.68)
	$B \to [KK]_D\pi$			$B \to [KK]_D K$		
	MC evts	*Absolute eff.(%)*	*Relative eff.(%)*	MC evts	*Absolute eff.(%)*	*Relative eff.(%)*
MC simulated	1034996	100.00	–	1007996	100.00	–
Preliminary	57167	(5.52 ± 0.02)	(5.52 ± 0.02)	56411	(5.60 ± 0.02)	(5.60 ± 0.02)
p and p_T cuts	48698	(4.71 ± 0.02)	(85.19 ± 0.53)	47538	(4.72 ± 0.02)	(84.27 ± 0.52)
BDT > 0.80	43056	(4.16 ± 0.02)	(88.41 ± 0.58)	41702	(4.14 ± 0.02)	(87.72 ± 0.59)
Charmonium veto	43056	(4.16 ± 0.02)	(100.00 ± 0.68)	41701	(4.14 ± 0.02)	(100.00 ± 0.69)
[*dau, bach*] Cross feed	42737	(4.13 ± 0.02)	(99.26 ± 0.68)	41374	(4.10 ± 0.02)	(99.22 ± 0.69)
$FD_{[z_D^{vtx}-z_B^{vtx}]} > 2$	33715	(3.26 ± 0.02)	(78.89 ± 0.57)	32547	(3.23 ± 0.02)	(78.67 ± 0.58)
$\lvert m_D - \mu_D \rvert <$ 25 MeV/c^2	33487	(3.24 ± 0.02)	(99.32 ± 0.77)	32284	(3.20 ± 0.02)	(99.19 ± 0.78)
Trigger	32079	(3.10 ± 0.02)	(95.80 ± 0.75)	30950	(3.07 ± 0.02)	(95.87 ± 0.76)
	$B \to [\pi\pi]_D\pi$			$B \to [\pi\pi]_D K$		
	MC evts	*Absolute eff.(%)*	*Relative eff.(%)*	MC evts	*Absolute eff.(%)*	*Relative eff.(%)*
MC simulated	1027494	100.00	–	1009996	100.00	–
Preliminary	58568	(5.70 ± 0.02)	(5.70 ± 0.02)	58146	(5.76 ± 0.02)	(5.76 ± 0.02)
p and p_T cuts	49636	(4.83 ± 0.02)	(84.75 ± 0.52)	48963	(4.85 ± 0.02)	(84.21 ± 0.52)
BDT > 0.80	44380	(4.32 ± 0.02)	(89.41 ± 0.58)	43710	(4.33 ± 0.02)	(89.27 ± 0.59)
Charmonium veto	44380	(4.32 ± 0.02)	(100.00 ± 0.67)	43709	(4.33 ± 0.02)	(100.00 ± 0.68)
[*dau, bach*] Cross feed	44080	(4.29 ± 0.02)	(99.32 ± 0.67)	43402	(4.30 ± 0.02)	(99.30 ± 0.67)

(continued)

Table 3.7 Continued

	$B \to [K\pi]_D\pi$			$B \to [K\pi]_D K$		
	MC evts	*Absolute eff.(%)*	*Relative eff.(%)*	MC evts	*Absolute eff.(%)*	*Relative eff.(%)*
$FD_{[z_D^{vtx} - z_B^{vtx}]} > 2$	35521	(3.46 ± 0.02)	(80.58 ± 0.57)	34829	(3.45 ± 0.02)	(80.25 ± 0.58)
$\|m_D - \mu_D\| <$ 25 MeV/c^2	34664	(3.37 ± 0.02)	(97.59 ± 0.74)	33979	(3.36 ± 0.02)	(97.56 ± 0.74)
Trigger	32911	(3.20 ± 0.02)	(94.94 ± 0.73)	32345	(3.20 ± 0.02)	(95.19 ± 0.74)

The first column of each table represents the total number of MC truth matched events which pass all the cuts. The second column is the absolute total efficiency calculated as $n_{\text{pass}}/n_{\text{generated}}$ events, while the third column is the relative efficiency of each cut after all the others are applied. The errors on the efficiencies are obtained propagating the pure statistical error on the number of generated events. The first set of cuts applied is described in Sect. 3.5, while background vetoes to remove charmonium, charmless and cross feed events are presented in detail in Chap. 4 but reported here for completeness. The chosen MVA discriminator cut is BDT > 0.92 for the favoured and suppressed modes and BDT > 0.80 for the *CP* modes. Favoured $\to$ suppressed cross feed veto (see Sect. 4.5.3) is applied only for the $[K\pi]_D$ and $[\pi K]_D$ samples. $\varepsilon_{\text{eff}}^{\text{TISTOS}}$ completes the selection (see Sect. 3.6)

Table 3.8 Expected $B \to [hh]_D h$ signal yield in the 2011 data sample used in the analysis

Mode	$\mathcal{B}_{B\to[hh]_D h}$	$\varepsilon_{\text{eff}}^{\text{PID}_{\text{dau}}}$	f_{corr}	Expected yield
$B \to [K\pi]_D\pi$	$(1.87 \pm 0.06) \times 10^{-4}$	0.702 ± 0.004	0.941 ± 0.008	~114505
$B \to [KK]_D\pi$	$(1.92 \pm 0.07) \times 10^{-5}$	0.831 ± 0.012	0.984 ± 0.008	~18805
$B \to [\pi\pi]_D\pi$	$(6.78 \pm 0.24) \times 10^{-6}$	0.649 ± 0.080	0.984 ± 0.008	~4921
$B \to [K\pi]_D K$	$(1.42 \pm 0.13) \times 10^{-5}$	0.702 ± 0.004	0.941 ± 0.008	~8757
$B \to [KK]_D K$	$(1.46 \pm 0.13) \times 10^{-6}$	0.831 ± 0.012	0.984 ± 0.008	~1448
$B \to [\pi\pi]_D K$	$(5.15 \pm 0.47) \times 10^{-7}$	0.649 ± 0.080	0.984 ± 0.008	~382
$B \to [\pi K]_D\pi$		$R_{\text{ADS}(\pi)} \sim 0.003$		~344
$B \to [\pi K]_D K$		$R_{\text{ADS}(K)} \sim 0.014$		~123

$$\text{PDF}_{\text{sig}}(x) = \exp\left(\frac{-(x-\mu)^2}{2\sigma^2 + \alpha_{L,R}(x-\mu)^2}\right) \tag{3.1}$$

where μ is the mean of the distribution, σ its width and α_L (for $\mu < 0$) and α_R (for $\mu > 0$) parameterise the non-Gaussian left and right tails.

3.9.2 D Mass Constraint and MC lineshapes

$B \to [hh]_D h$ candidates are refitted [14] constraining the vertices to points in space and the $[hh]_D$ candidates to the nominal D^0 mass [5]. The analysis takes considerable advantage of this refitting procedure since the widths of the signal peaks are reduced. The reduced width of the correctly reconstructed peak increases its separation from the irreducible $B \to Dh$ misidentified component, described in detail in Sect. 3.9.4. Additional benefits to the overall analysis strategy are the reduced dependence on the $[hh]_D$ particular final state, with momentum corrections to be applied only on the bachelor track (Sect. 3.10), and the smearing effect induced on the mass distributions of the $B \to hhh$ charmless modes (Sect. 4.4.2). The unconstrained D and B invariant mass distributions for signal MC simulated events are shown in Figs. 3.13 and 3.14, while the fitted widths of the peaks are summarized in Table 3.9.

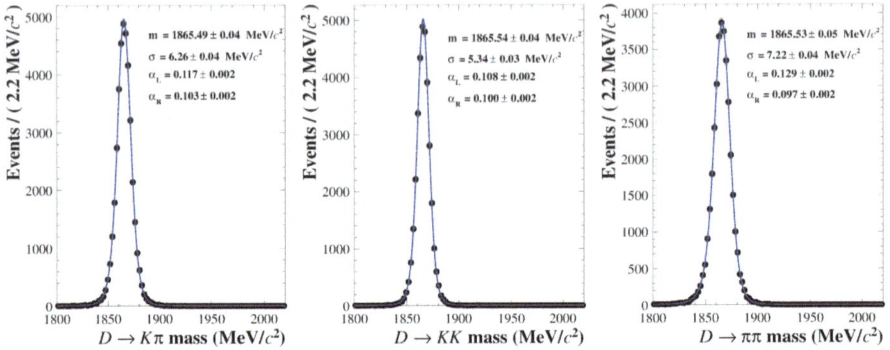

Fig. 3.13 **MC**: D invariant mass distributions for signal simulated, truth matched, events. No PID requirement applied on the daughter tracks

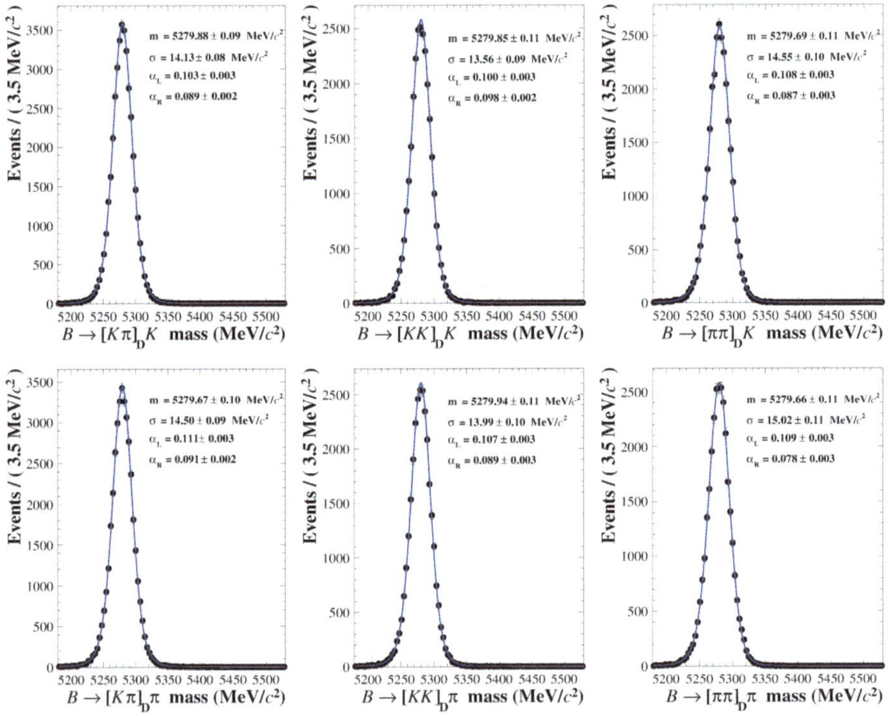

Fig. 3.14 **MC**: B invariant mass distributions for signal simulated, truth matched, events. No constraint is applied on the $[hh]_D$ candidate mass. (No PID requirement applied on both daughter and bachelor tracks)

Table 3.9 Widths of the signal mass peaks for the different MC simulated $B \to [hh]_D h$ modes (MeV/c^2) without applying the D mass constraint

	$[K\pi]_D$	$[KK]_D$	$[\pi\pi]_D$
$\sigma_{[hh]_D}$	6.26 ± 0.04	5.34 ± 0.03	7.22 ± 0.04
$\sigma_{[B \to DK]}$	14.13 ± 0.08	13.56 ± 0.09	14.55 ± 0.10
$\sigma_{[B \to D\pi]}$	14.50 ± 0.09	13.99 ± 0.10	15.02 ± 0.11
$\sigma_{DK}/\sigma_{D\pi}$	0.97 ± 0.01	0.97 ± 0.01	0.97 ± 0.01

The full offline selection is applied, but no requirement is applied on the PID $\Delta LL_{K/\pi}$ of both the bachelor and the daughter tracks

Table 3.10 Widths of the signal D-constrained, B mass peaks for the different MC simulated $B \to [hh]_D h$ modes (MeV/c^2)

	$[K\pi]_D$	$[KK]_D$	$[\pi\pi]_D$
$\sigma_{[hh]_D}$	–	–	–
$\sigma_{[B \to DK]}$	11.27 ± 0.06	11.12 ± 0.07	11.53 ± 0.08
$\sigma_{[B \to D\pi]}$	11.84 ± 0.07	11.86 ± 0.08	12.02 ± 0.08
$\sigma_{DK}/\sigma_{D\pi}$	0.95 ± 0.01	0.94 ± 0.01	0.96 ± 0.01

The full offline selection is applied, but no requirement is applied on the PID $\Delta LL_{K/\pi}$ of both the bachelor and the daughter tracks. The $[hh]_D$ invariant mass is constrained to the nominal D^0 mass value (delta-like functions with RMS < 1 KeV/c^2)

The resolution of both B and D peaks depends on the specific kaon and pion content of each final state. In fact, kaons carry on average less momentum than pions and the mass resolution is expected to improve the more kaons are in the $B \to [hh]_D h$ final state compared to pions. For example, in MC simulated events $\sigma_{[KK]_D}$ is about 15 % better than $\sigma_{[K\pi]_D}$, while $\sigma_{[\pi\pi]_D}$ is about 15 % worse. The same pattern applies to the B invariant mass with $B \to DK$ peaks being narrower than $B \to D\pi$ peaks. Once candidates are refitted with the D mass and vertex constraints, the widths of the peaks decrease on average of a factor of 20 %, as summarized in Table 3.10. Refitted mass distributions are shown in Figs. 3.15 and 3.16.

3.9.3 Lineshapes in Data and Signal Mass Windows

MC shapes are compared to data before and after D mass and vertex constraints. m_D distributions are shown in Fig. 3.17. Resolutions in data on the $[hh]_D$ peaks are found to be about 15 % bigger than those in MC simulated signal events:

$$\sigma_{[K\pi]_D} = 7.09 \pm 0.02 \quad \sigma_{[KK]_D} = 6.13 \pm 0.06 \quad \sigma_{[\pi\pi]_D} = 8.27 \pm 0.14$$

One can thus define a *signal D mass window* and apply a cut on the $[hh]_D$ invariant mass in order to further reject backgrounds which do not peak in the D mass

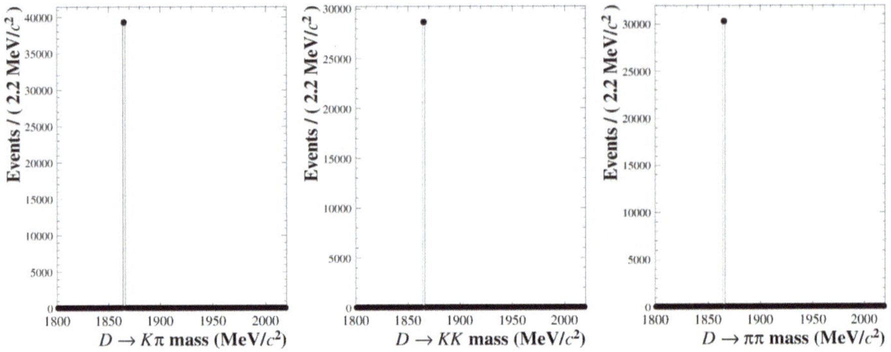

Fig. 3.15 **MC**: D invariant mass distributions for signal simulated, truth matched, events once the D mass is constrained to its nominal central value (approximate delta function with RMS < 1 KeV/c^2)

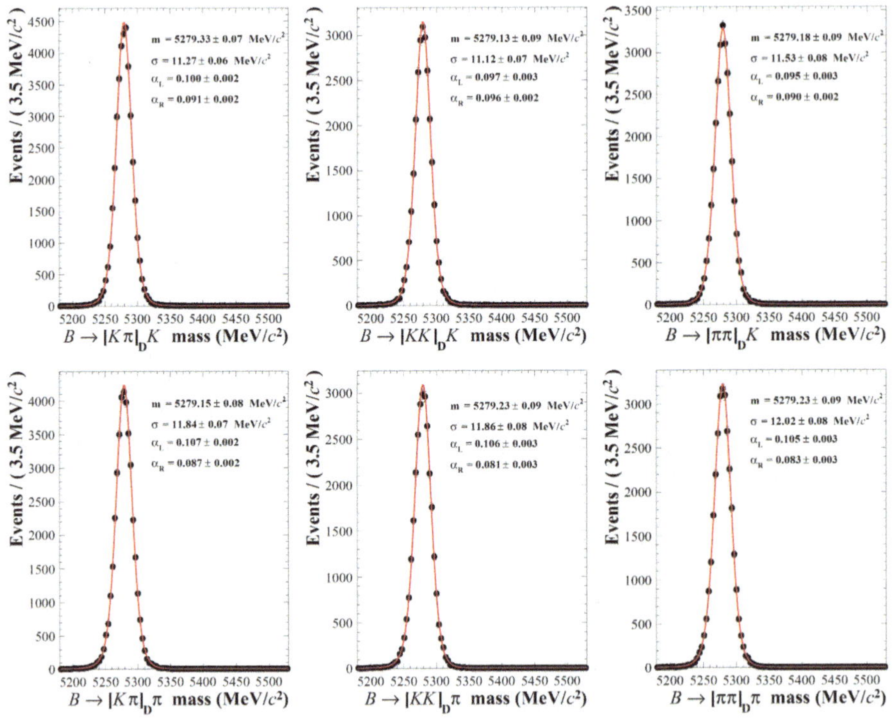

Fig. 3.16 **MC**: B invariant mass distributions for signal simulated, truth matched, events once the D mass is constrained to its nominal central value. (No PID requirement applied on both daughter and bachelor tracks)

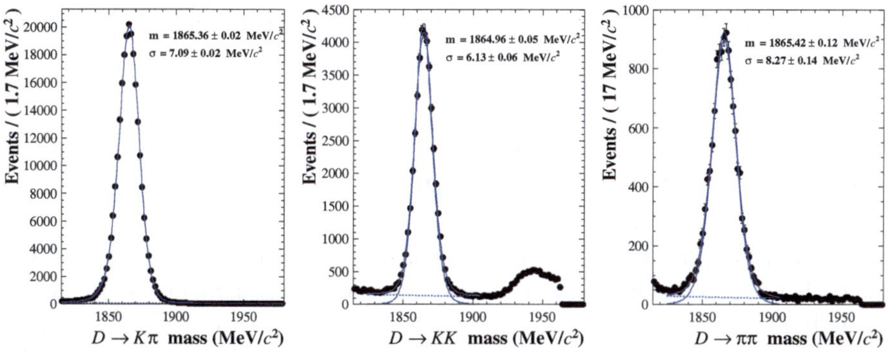

Fig. 3.17 **DATA**: D invariant mass distributions for $D \rightarrow hh$ offline selected events (PID requirements applied on the daughter tracks). Only the relevant range is fitted to exclude internal cross feed (Sect. 4.5.1)

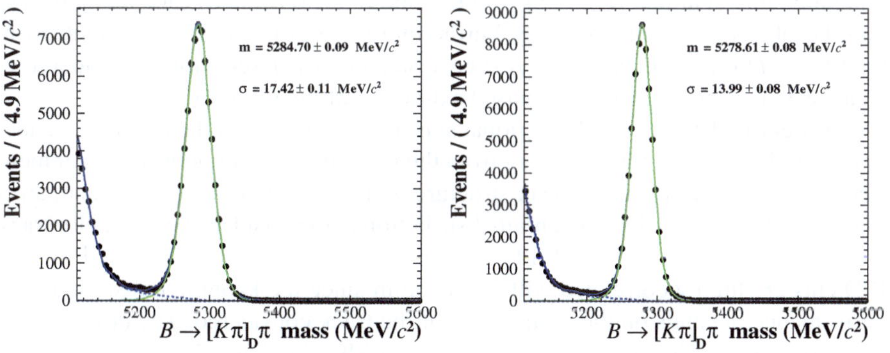

Fig. 3.18 **DATA**: B invariant mass distributions of the $B \rightarrow [K\pi]_D\pi$ control sample. *Left* B mass distribution before any mass constraint. *Right* B mass distribution after refitting the candidates with D mass and vertex constraints

distribution. The D signal window is defined as the $[-25, +25]\,\text{MeV}/c^2$ range around the center of the D peak, corresponding to about $3.5\,\sigma$.

The $B \rightarrow [K\pi]_D\pi$ favoured mode is then exploited to evaluate the effect of the D mass constraint in data and define an analogous B signal window. Although the full fit to the B mass distribution is performed in the wide range of $5100\text{–}5750\,\text{MeV}/c^2$ (Sect. 5.1), it is useful to define a *signal B mass window* for background study purpose (Sect. 4.3.2). The B invariant mass distribution of the offline selected data $B \rightarrow [K\pi]_D\pi$ favoured mode is shown in Fig. 3.18, before and after the refitting procedure. An expected 20 % improvement in the B peak resolution is found. The *signal B mass window* is thus defined as the $[-50, +50]\,\text{MeV}/c^2$ range around the center of the B peak, corresponding to about $3.5\,\sigma$ of the peak in data.

The MC/data differences in the resolution of the B peaks are used to smear all the background shapes described in Chap. 4 derived from MC simulated events. The smearing on the B mass distribution is performed adding a gaussian resolution effect of $\sigma_{\text{smear}} = \sqrt{\sigma^2_{\text{data}} - \sigma^2_{\text{MC}}} \sim 7\,\text{MeV}/c^2$.

3.9.4 $B \to Dh$ Misidentified Shape

This section describe the mass distribution of $B \to Dh$ genuine signal events where the bachelor track has been identified with the wrong mass hypothesis. This is particularly important for all the $B \to [hh]_D K$ signal modes, which suffer from the obvious pollution of the more abundant $B \to [hh]_D \pi$ modes where the bachelor pion is misidentified as a kaon. $B \to [hh]_D \pi$ and $B \to [hh]_D K$ decays are separated in the fitter via a PID cut on the bachelor track as described in Sect. 5.1. However, assuming a pion misidentification rate of order of $(1-5)\%$, the number of remaining misidentified $B \to D\pi$ events in the $B \to DK$ slice is not negligible and must be taken into account properly. Furthermore, since those events contribute to the total B $\to D\pi$ yield in each $[hh]_D$ slice, a satisfactory description of misidentified signal events is crucial for an unbiased extraction of the CP observables.

Misidentified events exhibit a global shift up (down) due to the artificially added (subtracted) invariant mass deriving from the incorrect mass hypothesis assignment. They have a broad displaced peak structure with a long tail on one side and, since this shape depends on the momentum distribution of the misidentified bachelor track, a dependence on any $\Delta LL^{bach}_{K/\pi}$ PID cut on the bachelor track is expected. In fact, PID cuts are highly correlated with momentum since the kaon/pion discrimination performance depends on the momentum of the track considered. The effects of the bachelor track misidentification are studied on MC simulated $B \to [hh]_D h$ signal events. Figure 3.19 shows the misidentified mass distributions as obtained with the true 4-momentum combination only of the signal tracks, in order to isolate the mass assignment contribution from the detector resolution effects. The Probability Density Functions (PDFs) to be used in the fitter are then shaped using the full detector simulation, as shown in Fig. 3.20. The shape of the $B \to [hh]_D \pi$ events, reconstructed as $B \to [hh]_D K$ is parameterised as the sum of two Crystal Ball functions, each defined as:

$$\text{CB}(x; \alpha, n, \mu, \sigma) = N \cdot \begin{cases} \exp\left(-\frac{(x-\mu)^2}{2\sigma^2}\right) & \text{for x} > -|\alpha| \\ \frac{\left(\frac{n}{|\alpha|}\right)^n e^{-\frac{\alpha^2}{2}}}{\left(\frac{n}{|\alpha|}-|\alpha|-x\right)^n} & \text{for x} < -|\alpha| \end{cases}$$

where N is the normalization factor, μ and σ the parameters of the core Gaussian part of the curve and α and n the power-law tail parameters. All parameters are taken from simulation and left floating in the fitter where possible. Shapes depend on the chosen PID cut on the bachelor track and the PID simulated performance on MC generated events is known not to describe data with good accuracy (a MC/data

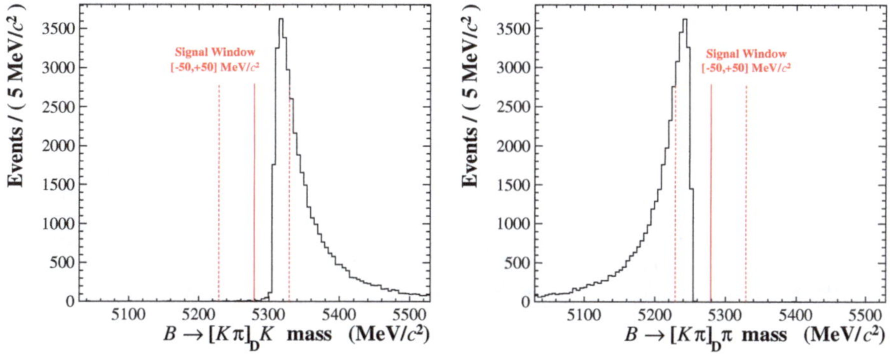

Fig. 3.19 *Left* $B \to [K\pi]_D\pi$ MC simulated events reconstructed with the $B \to [K\pi]_D K$ mass hypothesis (π misidentified as a K). *Right* $B \to [K\pi]_D K$ MC simulated events reconstructed with the $B \to [K\pi]_D\pi$ mass hypothesis (K misidentified as a π). The mass distributions are obtained using the combination of true 4-momenta only

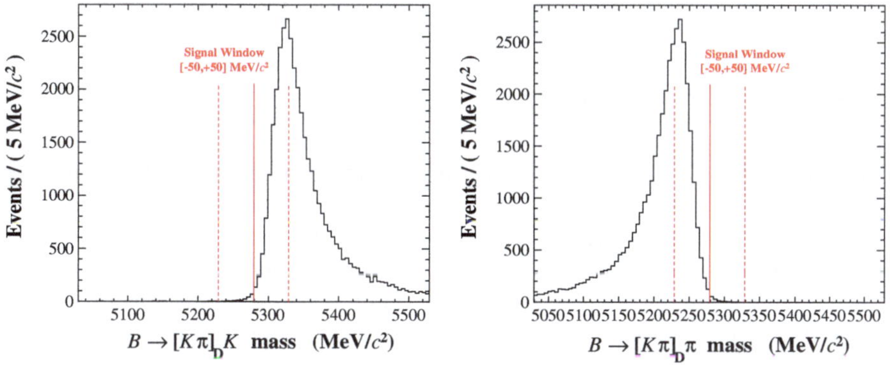

Fig. 3.20 *Left* $B \to [K\pi]_D\pi$ MC simulated events reconstructed with the $B \to [K\pi]_D K$ mass hypothesis (π misidentified as a K). *Right* $B \to [K\pi]_D K$ MC simulated events reconstructed with the $B \to [K\pi]_D\pi$ mass hypothesis (K misidentified as a π). Mass distributions are obtained using the full LHCb detector simulation

comparison for the PID response is shown briefly in Fig. 3.24). The general idea is thus to choose the most flexible PDF parameterisation as possible which ensures a good fitter stability and allows at the same time the most important parameters to adjust on data itself. Fixed and floating parameters in the fit are detailed in Sect. 5.1 and the chosen parameterisation has been validated with toy MC studies (Sect. 5.2). The same shape is used for all the $[hh]_D$ modes considered. The legitimacy of this strategy is justified from simulation: Figure 3.21 shows a simultaneous fit using the sum of two Crystal Balls, with the constraints used in the fitter, applied to all D modes.

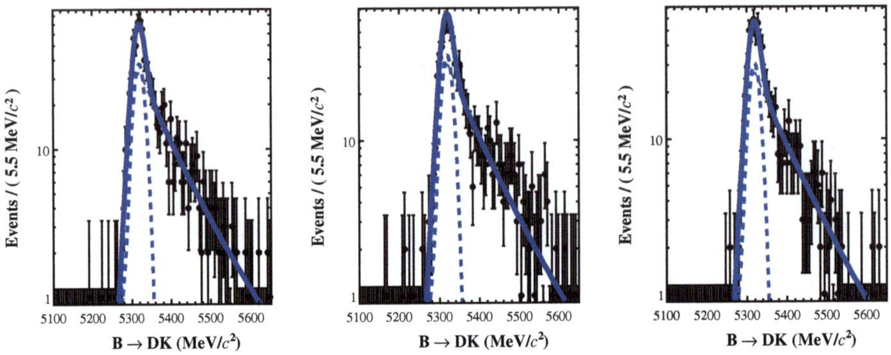

Fig. 3.21 Simultaneous fit to offline selected MC simulated events. $B \to [hh]_D\pi$ events are reconstructed with the $B \to [hh]_D K$ mass hypothesis. *Left* $[K\pi]_D$ mode. Center: $[KK]_D$ mode. *Right* $[\pi\pi]_D$ mode. A common PDF comprised of the sum of two Crystal Ball functions is fitted simultaneously to the three $[K\pi, KK, \pi\pi]_D$ submodes with a set of parameters in common. Good agreement is found in all modes. This allows the same shape to be used for all the $[hh]_D$ modes considered, since the tail is correctly described by the same set of parameters

3.10 Momentum Rescaling

Every $B \to [hh]_D h$ sample is split by charge and fitted to extract *CP* asymmetries. Signal peaks are fitted with the modified Gaussian PDF described in Sect. 3.9.1. When the favoured $B \to [K\pi]_D\pi$ sample is fitted separately splitting it by charge and polarity, the fitted means μ of the signal peaks are displaced from the established B mass value [5] as shown in Fig. 3.22. The observed shifts, defined as $\mu - m_{\text{PDG}}$, are summarized in Table 3.11 and depends on a residual misalignment of the LHCb detector.

The shifts are found to be different for B^- and B^+ candidates, with an expected left/right symmetry. $B^- \to Dh^-$ candidates in the magnet DOWN sample are in principle equivalent to $B^+ \to Dh^+$ candidates in the magnet UP sample as the bachelor track is bending in the same direction. The deviations of the $B^\pm$ fitted mass from the expected value of $5279.17 \pm 0.29\,\text{MeV}/c^2$ are converted into a momentum scale bias as follows. A complete description of the momentum scale correction procedure and the LHCb detector alignment can be found in [15, 16]. For a two-body decay $P \to d_1 d_2$, the invariant mass of the parent particle can be calculated experimentally as:

Table 3.11 Observed shifts $\mu - m_{\text{PDG}}$ of the fitted means of the favoured signal peaks

	$B^+ \to [K^+\pi^-]_D\pi^+$	$B^- \to [K^-\pi^+]_D\pi^-$
Magnet DOWN	$(4.46 \pm 0.32)\,\text{MeV}/c^2$	$(2.29 \pm 0.32)\,\text{MeV}/c^2$
Magnet UP	$(2.91 \pm 0.33)\,\text{MeV}/c^2$	$(4.59 \pm 0.33)\,\text{MeV}/c^2$

These shifts should be compared to the fitted peak width of $14.2\,\text{MeV}/c^2$

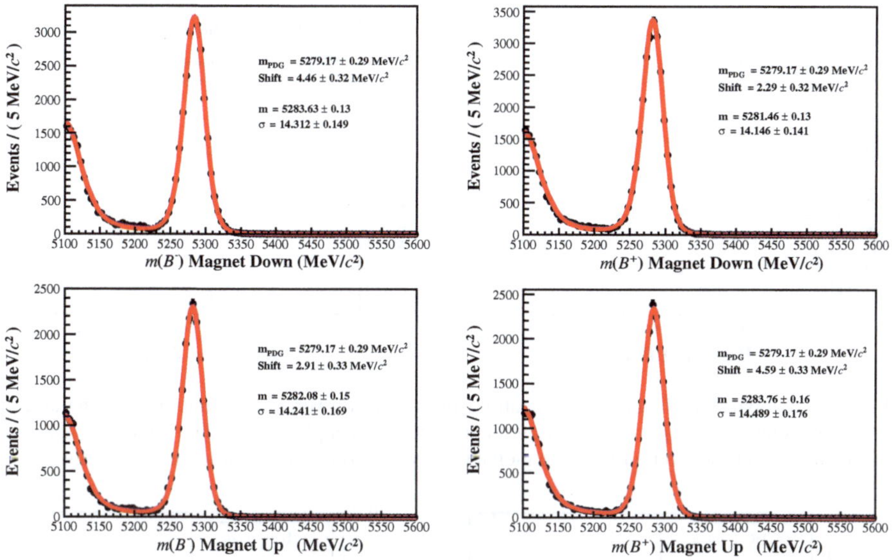

Fig. 3.22 Favoured $B \to [K\pi]_D\pi$ data sample split by magnet polarity and charge. Shifts in the means of the signal peaks are found

$$m_{12}^2 = (E_1 + E_2)^2 - (\vec{p}_1 + \vec{p}_2)^2$$

which can be simplified with a first order Taylor expansion in (m/p) [17] to:

$$\begin{gathered} m_{12}^2 = f + 2(p_1 p_2 - \vec{p}_1 \cdot \vec{p}_2) \\ \text{where} \\ f = m_1^2 + m_2^2 + \tfrac{p_2}{p_1} m_1^2 + \tfrac{p_1}{p_2} m_2^2 \end{gathered} \tag{3.2}$$

If the momentum scale is biased by a small factor α, the track vector momenta need to be multiplied by an overall factor $(1+\alpha)$ in order to obtain the *true* mass of the parent particle m_P. In the case that one of the daughter particles (d_2) is constrained to its nominal mass, only the momentum of the unconstrained particle p_1 must be rescaled properly as the momentum p_2 is assumed to be corrected by the mass constraint.[1] The unbiased mass of the decaying particle m_P can thus be expressed by:

$$\begin{gathered} m_P^2 = f' + 2(1+\alpha)(p_1 p_2 - \vec{p}_1 \cdot \vec{p}_2) \\ \text{where} \\ f' = m_1^2 + m_2^2 + \tfrac{p_2}{(1+\alpha)p_1} m_1^2 + \tfrac{(1+\alpha)p_1}{p_2} m_2^2 \end{gathered} \tag{3.3}$$

[1] This approximation assumes that the mass resolution is dominated by the momentum measurement of the daughter tracks and that the refitted quantities (the 4-momenta calculated after the mass constraint is applied) are used [17].

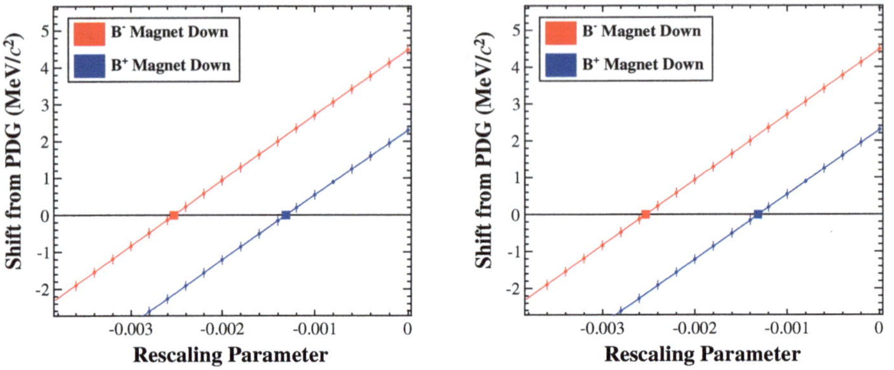

Fig. 3.23 Global shifts in the B mass central values for any given momentum rescaling parameter. Shifts are defined as $m_{\text{fitted}} - m_{\text{PDG}}$. *Left* magnet DOWN sample. *Right* magnet UP. The rescaling parameters are chosen to minimise the global observed shift in each sample independently

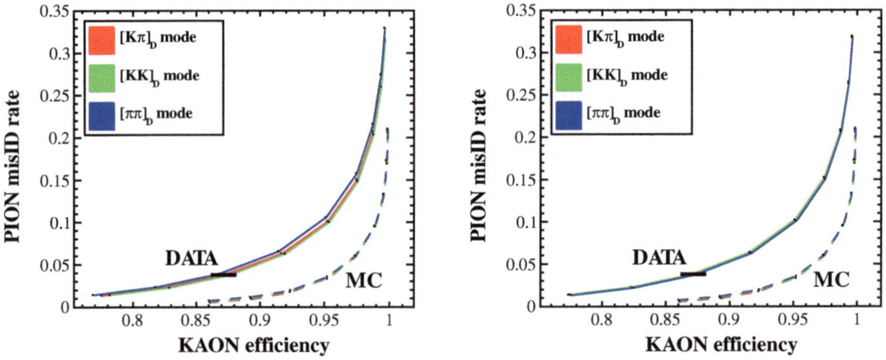

Fig. 3.24 ROC curve (ϵ_K^{PID} vs. $\epsilon_\pi^{\text{PID}}$) for kaon efficiencies and pion misidentification rates. *Left* magnet DOWN sample. *Right* magnet UP sample. Solid lines: data efficiencies taken from Tables 3.12, 3.13, 3.14 and 3.15. Dashed lines: simulated PID performance taken from signal MC generated events. The chosen working point corresponding to $\Delta LL_{K/\pi}^{bach} > 4$ is highlighted in black (with the assigned systematic errors as described in Sect. 5.3.2)

Taking the ratio of Eqs. 3.2 and 3.3:

$$\frac{m_{12}^2 - f}{m_P^2 - f'} = \frac{1}{(1+\alpha)}$$

which can be written, in the $\alpha \ll 1$ limit, as:

$$m_P^2 = (1+\alpha)m_{12}^2 - \alpha(m_1^2 + m_2^2) - 2m_1^2\alpha\frac{p_2}{p_1} \tag{3.4}$$

Equation 3.4 is used to rescale the measured invariant mass m_{12}^2 with any chosen α scale correction. For the $B \to [hh]_D h$ decays, the momentum of the bachelor track is thus rescaled by a factor determined on the most abundant $B \to [K\pi]_D\pi$ mode, such that data is centred at the known B mass central value. The best α parameters are chosen to minimise the $\mu - m_{\text{PDG}}$ shift for the favoured mode (separately for charge and magnet polarity) and the same scaling factors are applied to all the other $B \to [hh]_D h$ signal modes. This procedure is shown in Fig. 3.23 and allows to correct for the observed left/right μ asymmetries. Other benefits include that a single PDF can be used for all the $B \to [hh]_D h$ slices in the fitter (Sect. 5.1.1) removing the need to have different shifted background shapes for different charges and dipole polarities.

3.11 PID Calibration

Since $B \to [hh]_D K$ and $B \to [hh]_D\pi$ events are distinguished using the $\Delta LL_{K/\pi}^{bach}$ PID variable, as described in Sect. 5.1, a precise estimation of the efficiencies for each given PID cut on the bachelor track is needed. These efficiencies set the amount of misidentified $B \to D\pi$ events remaining after a cut on the PID variable. The PID calibration is performed with a totally data-driven procedure, exploiting a series of high statistics modes that can be reconstructed cleanly with kinematical cuts only [18]. A data-driven procedure is essential in this respect since PID variables are generally not well described in MC simulation, which assumes a perfect detector with no magnetic distortion effects, no temperature variations in the radiator and other non-linear effects. The method for calibrating the performance of the RICH detectors in identifying kaons and pions uses a high-statistics $D^{*+} \to D^0(K^-\pi^+)\pi^+$ calibration sample (charge conjugation implied). The calibration sample is taken from the full 2011 LHCb dataset. In this decay both the daughter kaon and pion tracks of the D^0 meson are *self-tagged* by the relative charge of the bachelor pion from the D^* resonance, thus providing an independent high purity sample of kaon and pion tracks selected without applying any PID requirement.[2] Since PID performance depends on several factors, i.e. the momentum p of the track and its pseudo-rapidity η, the calibration sample is reweighted such that its spectra follow those of the signal. Each sample is binned into $18 \times 4 \times 4$ bins in $|\vec{p}|$, η and number of tracks respectively, in order to take into account the different RICH response, e.g. high occupancy events reduce the RICH overall performance. For each bin, the PID performance is mapped in the calibration tracks and the total, effective performance of each $\Delta LL_{K/\pi}^{bach}$ cut on our sample is deduced from the weighted sum of the $D^{*\pm}$ calibration-sample's performance in each bin. The results of this procedure is shown in Tables 3.12 and 3.13 for the kaon efficiencies ϵ_K^{PID} while Tables 3.14 and 3.15 show the corresponding pion misidentification rates $\epsilon_\pi^{\text{PID}}$ for a range of the bachelor's $\Delta LL_{K/\pi}^{bach}$ cut. The

[2] The double-Cabibbo suppressed $D^0(K^+\pi^-)$ mode is vetoed in the D^* sample selection [18].

Table 3.12 $\Delta LL_{K/\pi}^{bach}$ PID efficiencies for KAON tracks, magnet DOWN sample

$\Delta LL_{K/\pi}^{bach} >$	$\epsilon_{[K\pi]_D}$	$\sigma_{[K\pi]_D}$	$\epsilon_{[\pi\pi]_D}$	$\sigma_{[\pi\pi]_D}$	$\epsilon_{[KK]_D}$	$\sigma_{[KK]_D}$
8.0	0.775826	0.000176	0.768838	0.000178	0.782170	0.000172
6.0	0.823795	0.000163	0.818359	0.000166	0.829420	0.000159
4.0	0.872001	0.000146	0.868265	0.000149	0.876284	0.000142
2.0	0.917125	0.000124	0.914620	0.000127	0.919691	0.000121
0.0	0.952690	0.000101	0.951369	0.000103	0.953939	0.000099
−2.0	0.975612	0.000080	0.975169	0.000082	0.976087	0.000079
−4.0	0.987953	0.000064	0.987718	0.000067	0.988032	0.000065
−6.0	0.993940	0.000054	0.993798	0.000056	0.993899	0.000055
−8.0	0.996720	0.000048	0.996619	0.000050	0.996662	0.000049

Table 3.13 $\Delta LL_{K/\pi}^{bach}$ PID efficiencies for KAON tracks, magnet UP sample

$\Delta LL_{K/\pi}^{bach} >$	$\epsilon_{[K\pi]_D}$	$\sigma_{[K\pi]_D}$	$\epsilon_{[\pi\pi]_D}$	$\sigma_{[\pi\pi]_D}$	$\epsilon_{[KK]_D}$	$\sigma_{[KK]_D}$
8.0	0.774093	0.000213	0.776625	0.000213	0.775154	0.000217
6.0	0.822631	0.000198	0.825236	0.000198	0.823573	0.000201
4.0	0.871097	0.000177	0.873410	0.000177	0.871728	0.000181
2.0	0.915989	0.000151	0.917771	0.000152	0.916142	0.000155
0.0	0.951685	0.000123	0.952599	0.000124	0.951564	0.000127
−2.0	0.974943	0.000098	0.975187	0.000100	0.974594	0.000102
−4.0	0.987627	0.000080	0.987589	0.000082	0.987229	0.000083
−6.0	0.993648	0.000068	0.993575	0.000070	0.993468	0.000070
−8.0	0.996514	0.000060	0.996408	0.000062	0.996420	0.000062

uncertainties in the tables are statistical errors from the calibration sample only as they depend purely on the size of the $D^{*\pm}$ sample. A systematic error is then assigned to take into account the fact that the binning choice may not perfectly parameterise the variation in the RICH performance, as described in Sect. 5.3.2. The ROC curve (ϵ_K^{PID} v.s. $\epsilon_\pi^{\text{PID}}$) for both magnet polarities is shown in Fig. 3.24. All efficiencies are quoted for the $[K\pi, KK, \pi\pi]_D$ different submodes depending on the different $B \to Dh$ signal mode used as a target reference.

Eventually, in the final selection a cut of 4 is chosen and both polarity datasets are added together. Therefore, the PID efficiency is taken as

$$\epsilon_\pi^{\text{PID}} = (3.8 \pm 0.2)\% \quad \epsilon_K^{\text{PID}} = (87.2 \pm 1.0)\%$$

where the quoted error is the assigned systematic uncertainty (see Sect. 5.3.2).

Note that the same calibration procedure is not performed for the two D daughter tracks, since the same requirement is applied to all modes ($\Delta LL_{K/\pi}^{dau} > 2$ for the daughter kaons and $\Delta LL_{K/\pi}^{dau} < -2$ for the daughter pions) and only relative mesurements are extracted in the analysis.

Table 3.14 $\Delta LL^{bach}_{K/\pi}$ misidentification rates for PION tracks, magnet DOWN sample

$\Delta LL^{bach}_{K/\pi} >$	$\epsilon_{[K\pi]_D}$	$\sigma_{[K\pi]_D}$	$\epsilon_{[\pi\pi]_D}$	$\sigma_{[\pi\pi]_D}$	$\epsilon_{[KK]_D}$	$\sigma_{[KK]_D}$
8.0	0.013910	0.000053	0.014466	0.000053	0.014118	0.000052
6.0	0.022734	0.000065	0.023682	0.000066	0.022873	0.000065
4.0	0.037685	0.000082	0.039275	0.000083	0.037715	0.000081
2.0	0.06301	0.000103	0.065906	0.000105	0.062569	0.000102
0.0	0.101113	0.000128	0.106056	0.000130	0.100082	0.000125
−2.0	0.151065	0.000152	0.158116	0.000154	0.148893	0.000148
−4.0	0.207121	0.000171	0.21632	0.000172	0.203736	0.000167
−6.0	0.26369	0.000186	0.274773	0.000185	0.259672	0.000181
−8.0	0.317631	0.000195	0.330031	0.000193	0.313364	0.000191

Table 3.15 $\Delta LL^{bach}_{K/\pi}$ misidentification rates for PION tracks, magnet UP sample

$\Delta LL^{bach}_{K/\pi} >$	$\epsilon_{[K\pi]_D}$	$\sigma_{[K\pi]_D}$	$\epsilon_{[\pi\pi]_D}$	$\sigma_{[\pi\pi]_D}$	$\epsilon_{[KK]_D}$	$\sigma_{[KK]_D}$
8.0	0.013910	0.000053	0.014466	0.000053	0.014118	0.000052
6.0	0.022734	0.000065	0.023682	0.000066	0.022873	0.000065
4.0	0.037685	0.000082	0.039275	0.000083	0.037715	0.000081
2.0	0.06301	0.000103	0.065906	0.000105	0.062569	0.000102
0.0	0.101113	0.000128	0.106056	0.000130	0.100082	0.000125
−2.0	0.151065	0.000152	0.158116	0.000154	0.148893	0.000148
−4.0	0.207121	0.000171	0.21632	0.000172	0.203736	0.000167
−6.0	0.26369	0.000186	0.274773	0.000185	0.259672	0.000181
−8.0	0.317631	0.000195	0.330031	0.000193	0.313364	0.000191

References

1. The LHCb Collaboration, Absolute luminosity measurements with the LHCb detector at the LHC. J. Instrum. **7**(01), P01010 (2012)
2. T. Sjöstrand, S. Mrenna, P. Skands, PYTHIA 6.4 physics and manual. JHEP **05**, 026 (2006) http://arXiv:hep-ph/0603175
3. D.J. Lange, The EvtGen particle decay simulation package. Nucl. Instrum. Meth. **A462**, 152 (2001)
4. GEANT4 collaboration, S. Agostinelli et al., GEANT4: a simulation toolkit. Nucl. Instrum. Meth. **A506**, 250 (2003)
5. Particle Data Group, K. Nakamura et al., Review of particle physics. J. Phys. **G37**, 075021 (2010)
6. LHCb collaboration, A measurement of the ratio of branching fractions: $(B^{\pm} \rightarrow DK^{\pm})/(B^{\pm} \rightarrow D\pi^{\pm})$ for $D \rightarrow K\pi$, KK, $K\pi\pi\pi$ and $K^0_S\pi\pi$, LHCb CONF-2011-031 (2011), LHCb-ANA-2010-013; LHCb-ANA-2011-053
7. A. Hoecker, P. Speckmayer, J. Stelzer, J. Therhaag, E. von Toerne, H. Voss, TMVA: toolkit for multivariate data analysis. PoS ACAT 040 (2007) http://arXiv:physics/0703039
8. CDF collaboration, T. Aaltonen et al., Measurements of branching fraction ratios and CP-asymmetries in suppressed $B^- \rightarrow D(\rightarrow K^+\pi^-)K^-$ and $B^- \rightarrow D(\rightarrow K^+\pi^-)\pi^-$ decays. Phys. Rev. **D84**, 091504 (2011) http://arXiv:1108.5765
9. L. Breiman, J. Friedman, R. Olshen, C. Stone, *Classification and Regression Trees* (Wadsworth, Belmont, CA, 1984)

10. LHCb Collaboration, R. Aaij et al., Measurement of $\sigma(pp \to b\bar{b}X)$ at $\sqrt{s} = 7$ TeV in the forward region. Phys. Lett. **B694**, 209 (2010) http://arXiv:1009.2731
11. CDF Collaboration, T. Aaltonenet al., Measurement of ratios of fragmentation fractions for bottom hadrons in $p\overline{p}$ collisions at $\sqrt{s} = 1.96$ TeV, Phys. Rev. D **77**, 072003 (2008)
12. The LHCb Collaboration, R. Aaji et al., Measurement of b hadron production fractions in 7 TeV pp collisions. Phys. Rev. D **85**, 032008 (2012)
13. Heavy Flavor Averaging Group, Y. Amhis et al., *Averages of B-Hadron, C-Hadron, and Tau-Lepton Properties as of Early 2012*, http://arXiv:1207.1158
14. Hulsbergen, W. D., Decay chain fitting with a Kalman filter. Nucl. Instrum. Meth. Phys. Res. A **552**, 566 (2005)http://arXiv:physics/0503191
15. R. Aaij et al., Measurement of b-hadron masses. Phys. Lett. B **708**(35), 241 (2012)
16. Y. Amhis, J. He, R. Mrki, M. Needham, O. Schneider, Study of the 2010 alignment stability using J/ψ and D^0 decays, Tech. Rep.LHCb-INT-2011-013. CERN-LHCb-INT-2011-013, CERN, Geneva, April 2011
17. M. Needham, Momentum scale calibration using resonances, Tech. Rep. LHCb-2008-037. CERN-LHCb-2008-037. LPHE-2008-08, CERN, Geneva, July 2008
18. https://twiki.cern.ch/twiki/bin/view/LHCb/PIDCalibPackage

Chapter 4
Background Studies

This chapter discusses possible sources of background for the eight $B \to [hh]_D h$ modes. This study uses both simulated events (later referred to as MC events) and, where possible data, in order to understand the background rates and their shapes in the invariant mass distribution of the $B^{\pm}$ candidates. The chapter starts with an introduction of the combinatorial background, this is presented in Sect. 4.1. Section 4.2 is dedicated to the description of modes involving muons in the final state, e.g. semileptonic or charmonium decays, which have high branching fractions and can potentially give contributions in the final sample. Most of the background comes from partially reconstructed events where one or more particles is lost in reconstruction. This kind of background, described in Sect. 4.3, is responsible for structures in the region immediately below the nominal B mass peak due to the missing energy associated with the non reconstructed particle. Two other important sources of peaking backgrounds like possible internal cross feed between the different signal modes considered and non resonant charmless events are considered in Sects. 4.5 and 4.4 respectively. $B \to D\pi$ events where the bachelor pion has been misidentified as a kaon are considered as part of the signal and described separately in Sect. 3.9. Every background relevant to the analysis, where stated, has been shaped in the region $m_B > 5000\,\text{MeV/c}^2$ to be included in the signal fitter (Sect. 5.1).

4.1 Combinatorial Background

Combinatorial background, e.g. formed by random combinations of charged tracks coming from the event, is rejected by the multivariate selection presented in Sect. 3.5. Although highly suppressed by the cut on the BDT variable, some events remain in the final sample and must be considered properly. Two possible sources are identified depending on whether the D^0 candidate is associated with a true D^0 meson or not. In the former case, the event has only a single displaced vertex associated with the D^0 meson (prompt charm production, not from a b quark, or a high flight distance D^0

P. Gandini, *Observation of CP Violation in $B^{\pm} \to DK^{\pm}$ Decays*,
Springer Theses, DOI: 10.1007/978-3-319-01029-8_4,

meson from other physical processes), while the latter is just a random combination of three tracks deriving from the primary vertex interaction and the rest of the event. The remaining combinatorial background in the final sample is distributed evenly in the whole mass distribution and is taken into account, with a linear PDF in the fit with floating yield and slope parameter.

Random Combinations of Three Tracks

The level of combinatorial background is checked by reconstructing data with the $B^{\pm} \to [h^{\pm}h^{\pm}]_D h^{\pm}$ hypothesis, with the sum of the electrical charges being equal to ± 3. In this way, one is able to obtain a pure, independent sample of random combinations of tracks with characteristics identical to those of the genuine signal events. The superposition of fake $B^{\pm} \to [K^{\pm}\pi^{\pm}]_D h^{\pm}$ and genuine $B^{\pm} \to [K^{\pm}\pi^{\mp}]_D h^{\pm}$ candidates after reconstruction and before offline selection, normalized to the same luminosity, is shown in Fig. 4.1 for the $35.7\,\mathrm{pb}^{-1}$ 2010 dataset used in the selection training and validation. After applying the BDT cut, only few events are left in the $B^{\pm} \to [K^{\pm}\pi^{\pm}]_D h^{\pm}$ sample, showing that the chances to imitate two displaced vertices for both the B and the D candidates are limited.

True D^0 Meson and Fake Bachelor

Combinatorial background can potentially arise also from the combination of a true $D^0 \to K\pi$ meson and a bachelor track coming from the rest of the event, typically

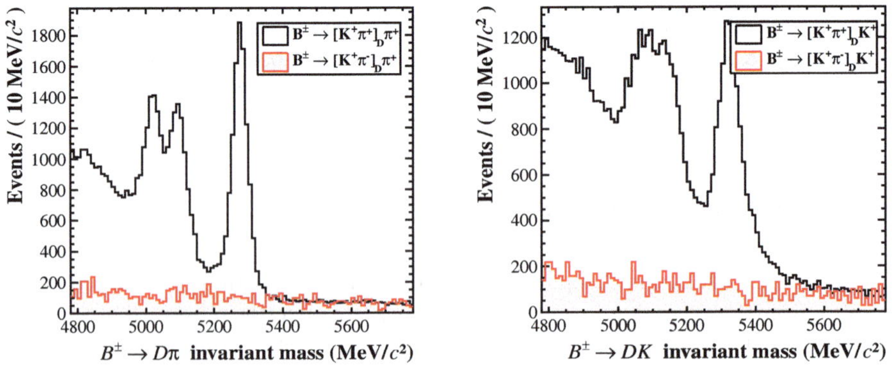

Fig. 4.1 Mass distributions for $B \to [K\pi]_D\pi$ (*left*) and $B \to [K\pi]_D K$ (*right*) after reconstruction and no offline selection applied. The *solid line* represents the $B \to Dh$ candidates with $[K^{\pm}\pi^{\mp}]_D$ while the *shaded area* represents candidates with $[K^{\pm}\pi^{\pm}]_D$. Plots are normalized to a luminosity of $35.7\,\mathrm{pb}^{-1}$

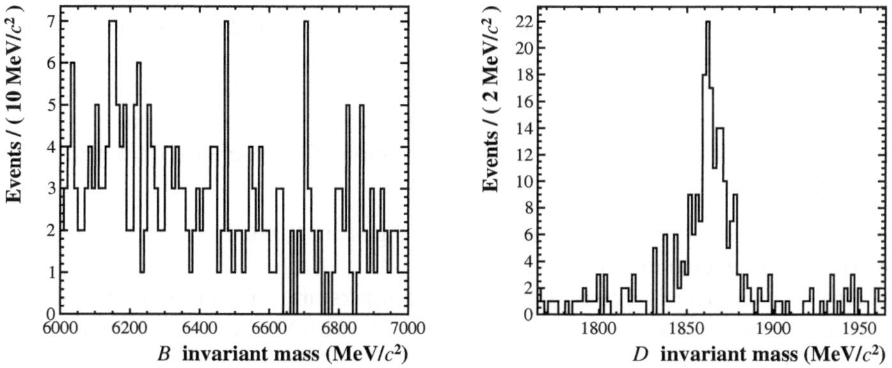

Fig. 4.2 B and D mass distributions for $B \to [K\pi]_D\pi$ data with $m_B > 6000\,\text{MeV/c}^2$ (luminosity $1.0\,\text{fb}^{-1}$). A clear contribution from prompt charm is visible

from the same primary vertex. In this case, only one true displaced vertex is present (B or D) in the reconstructed decay chain and the B candidate has a relatively short flying distance (especially when the D^0 comes directly from the primary interaction). The presence of true charm events can be investigated looking at the region $m_B > 6000\,\text{MeV/c}^2$, well above the expected $B^\pm$ nominal mass. The mass distributions for both B and D candidates in the considered case are shown in Fig. 4.2.

4.2 Backgrounds from Decays with Muons

This section discusses backgrounds containing one or more muons in the final state. The presence of decays involving muons as a source of background to a purely hadronic mode like $B \to [hh]_D h$ is due to the high trigger efficiency for muons in LHCb (Sect. 2.2.8). These kind of events are rejected using trigger requirements for the signal candidates as described in Sect. 3.6, e.g. vetoing events selected by muon triggers, as well as applying specific invariant mass vetoes in the selection. In the LHCb experiment, tracks are associated to the muon hypothesis by requiring that the track has hits in the muon chambers, as presented in Sect. 2.2.10. Hadrons are expected to be blocked by the calorimeters although part of the signal is expected to penetrate the hadronic calorimeter and give hits in the muon chambers (*punch-through* signal events where at least one hadron in the decay chain is not completely absorbed by the calorimeters).

4.2.1 Charmonium Veto

A particular source of background comes from charmonium events including $J/\psi \to \mu^+\mu^-$ and $\psi(2S) \to \mu^+\mu^-$ resonances. Modes like $B^\pm \to J/\psi K^\pm$ have the same number of tracks in the final state and peak close to the B mass when the two muons are misidentified as pions. These modes have high branching fractions of $\mathcal{O}(10^{-3})$ [1] and can survive into the offline selected sample even after the muon trigger requirements. In order to remove completely these charmonium events, a mass veto is introduced on the neutral combination of the bachelor track and one of the daughter tracks of the D meson as follows:

- Combinations of two tracks with opposite signs are formed
- Tracks are given a muon mass hypothesis
- Any B candidate is vetoed if both of the tracks are flagged as muons and the candidate's dimuon invariant mass lies within 55 MeV/c^2 of the nominal mass of either J/ψ or $\psi(2S)$ resonances [1].

The reconstructed charmonium invariant mass for data candidates is shown in Fig. 4.3, with and without the muon requirement described above. The veto has been tested both on a MC simulated sample of $B^{\pm} \rightarrow J/\psi K^{\pm}$ events reconstructed as

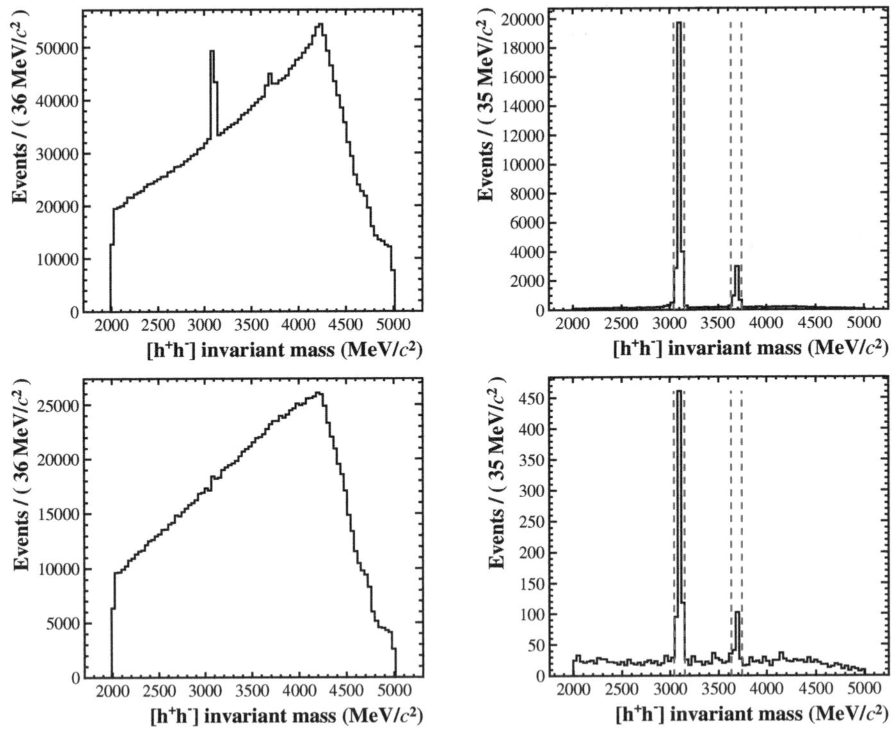

Fig. 4.3 Dimuon invariant mass of the $[h^+h^-]$ pairs for $B \rightarrow [K\pi]_D\pi$ data candidates after reconstruction (*top plots*) and after hadronic trigger requirements (*bottom plots*). *Left* combinations without any muon association requirement, *Right* muon association is required on both tracks forming the dimuon invariant mass. A contribution from both J/ψ and $\psi(2S)$ resonances is clearly visible along with the chosen veto window of $\{+55, -55\}$ MeV/c^2 ($\sim 3\sigma$ of the resonance peak)

$B \to [K\pi]_D\pi$ and on the signal MC simulated samples. Over about 1.5 million reconstructed events, the veto removes more than 100.0 % of offline-selected J/ψ events, while retaining 100.0 % of the signal decay.

4.2.2 Semileptonic Backgrounds

Background can arise from semileptonic $B^{\pm}$ decays, like $B \to D^0\mu\nu$ where the D^0 meson decays into a correctly reconstructed D final state and the muon is misidentified as a pion. These modes have high branching fractions (e.g. $\mathcal{B}(B^+ \to D^0 l^+ \nu_l) = (2.23 \pm 0.11) \times 10^{-2}$ [1]) although their contribution is sizable only when the neutrino is particularly soft and the reconstructed mass $m_{D\mu}$, formed using only the charged tracks and neglecting the neutrino, is close to the B meson mass. This process has been studied using both exclusive MC simulations and data.
$B^{\pm} \to [K\pi]_D\mu^{\pm}\nu$ decays have been simulated requiring that all the three charged tracks lie within the LHCb detector acceptance (no requirement on the neutrino). The full Dalitz plot of this decay, using the true 4-momenta of all the generated particles, is presented in Fig. 4.4 along with its projection on the $D\mu$ *reconstructible* mass. The *reconstructible* mass of the decay, defined as the invariant mass of the charged tracks excluding the neutrino, is a broad distribution peaking at about 3600 MeV/c^2, with a well defined kinematic edge at the mass of the B meson corresponding to the limit-configuration where the neutrino is produced at rest. Events with a reconstructible mass >5000 MeV/c^2, the region most relevant to the definition of the PDFs used in the fitter, have been generated with the full LHCb detector simulation. These are shown in Fig. 4.5 for different mass hypotheses of the bachelor track (with the true muon misidentified as the bachelor hadron).

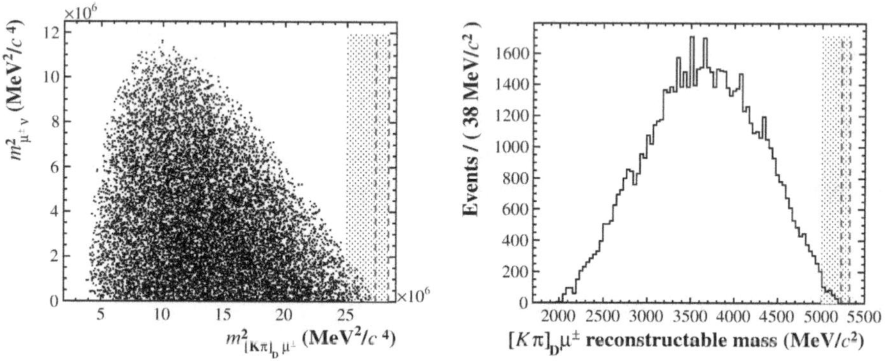

Fig. 4.4 $B^{\pm} \to [K\pi]_D\mu^{\pm}\nu$ MC simulated events. Events have been generated using the ISGW2 model in the EvtGen package [2]. *Left* full Dalitz plot of the three-body semileptonic decay. *Right* distribution of the $[K\pi]_D\mu$ invariant mass. *Dotted lines* represent the nominal B signal window, while the *shaded area* indicates the region >5000 MeV/c^2

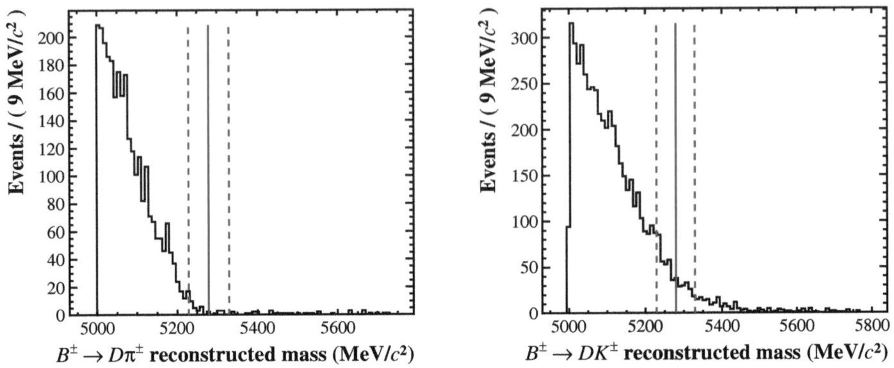

Fig. 4.5 $B^\pm \to [K\pi]_D\mu^\pm\nu$ events simulated with the full LHCb detector simulation. *Left* $B \to [K\pi]_D\pi$ hypothesis. *Right* $B \to [K\pi]_D K$ hypothesis. *Dotted lines* represent the nominal B signal window. These events, once properly weighted, are used within the partially reconstructed PDF described in Sect. 4.3

In order to quantify semileptonic decays in the offline selected sample on a purely data-driven basis, data is selected with and without applying the LHCb muon requirement described in Sect. 2.2.10. Data with and without this requirement is shown in Fig. 4.6 and shapes are in good agreement with the MC simulations presented in Fig. 4.5. Events with the bachelor track consistent with the muon hypothesis from the associated hits in the muon chambers appear to be less than 5 % the total yield (after hadronic trigger requirements) of the partially reconstructed backgrounds which populate the region >5000 MeV/c^2. For this reason, semileptonic backgrounds are not included in the fitter as a separate independent component but they are embedded in the partially reconstructed PDF, presented in detail in Sect. 4.3.
Using a big sample of simulated inclusive semileptonic $B \to Xl\nu_l$ decays, one can also identify other semileptonic modes not taken into account so far. Small contributions are found from $B \to D^*\mu\nu$ and $B \to D^{**}\mu\nu$ decays, where the higher-excited D^* resonances decay in chain into D^0 mesons and other neutral particles which are not reconstructed. For these modes, all the studies presented above apply in the exact same way although the missing mass, due to the extra neutral particles lost, shifts their kinematic endpoints well below 5000 MeV/c^2. For this reason they have been neglected in this section.

4.3 Partially Reconstructed Backgrounds

A significant source of background in the final sample is represented by decays with topologies similar to those of the signal, but with one or more missing particles that have not been reconstructed. Since one or more 4-momentum is missing in the computation of the invariant mass of the B meson, these events are distributed in the

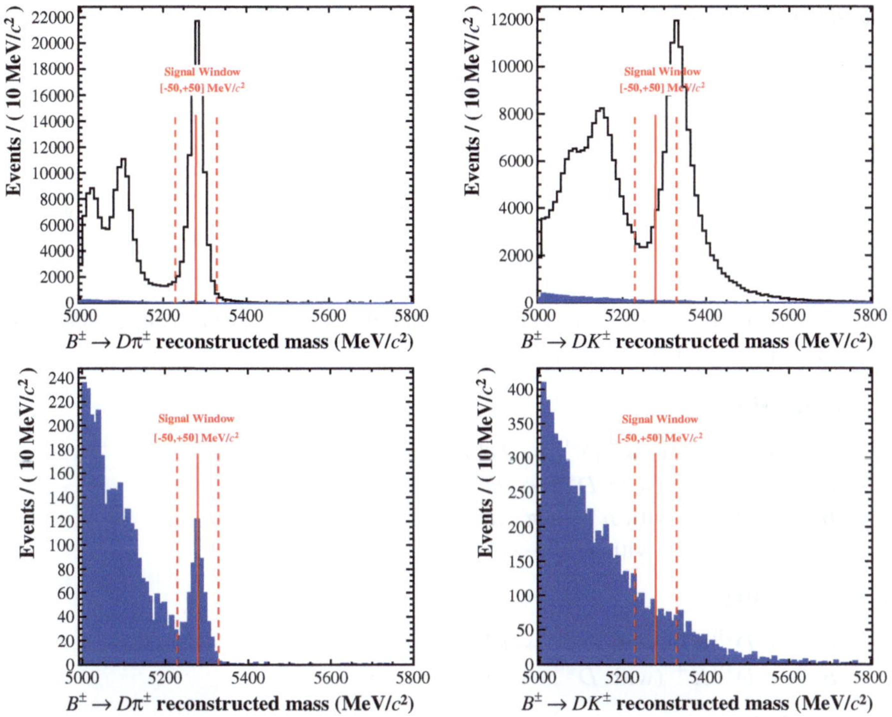

Fig. 4.6 $B \to [K\pi]_D h$ data reconstructed as $B \to [K\pi]_D\pi$ (*left*) and $B \to [K\pi]_D K$ (*right*) without any PID requirement on the bachelor track. The fraction of candidates with the bachelor track flagged as muon is shown in solid colour in both plots (the *bottom plot* presents a zoomed version to highlight the shape of the selected semileptonic background). The small peak, centered on the B meson mass, is caused by events in which a genuine bachelor pion track has been flagged as a muon (*punch-through* events). These events are removed with the charmonium veto

region immediately below the nominal B mass peak. These decay channels usually involve neutral or charged D^* resonances decaying into correctly reconstructed D^0 mesons and have high, well-known branching fractions of order of $\mathcal{O}(10^{-2})$; they are also responsible for typically visible structures on data, like double peaks in the region 5000–5200 MeV/c^2 due to the particular spin-angular dependence in the decay chain.

4.3.1 Partially Reconstructed Topology

In order to understand the properties of partially reconstructed backgrounds and how they distribute on the B mass spectrum, one can initially focus on those modes where only one particle is lost in reconstruction leading to missing invariant mass. The typical topology of those kind of events is sketched in Fig. 4.7. Several exclusive

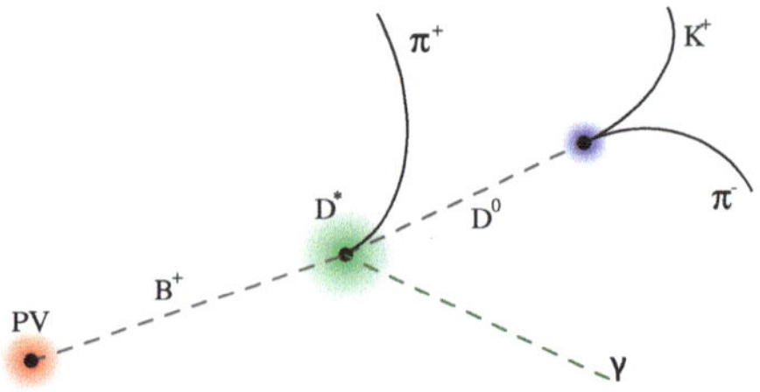

Fig. 4.7 Topology of $B \to D^*\pi$ decays, with $D^* \to D^0\gamma$

modes (10 in total) have been investigated in detail, and they can be grouped according to the non reconstructed particle:

- **One missing** π^0
  - $B^+ \to D^{*0}\pi^+$ (with $D^{*0} \to D^0\pi^0$)
  - $B^+ \to D^{*0}K^+$ (with $D^{*0} \to D^0\pi^0$)
  - $B^+ \to D^0\rho^+$ (with $\rho^+ \to \pi^+\pi^0$)
  - $B^+ \to D^0K^{*+}$ (with $K^{*+} \to K^+\pi^0$)
- **One missing** γ
  - $B^+ \to D^{*0}\pi^+$ (with $D^{*0} \to D^0\gamma$)
  - $B^+ \to D^{*0}K^+$ (with $D^{*0} \to D^0\gamma$)
- **One missing** $\pi^\pm$
  - $B^0 \to D^{*-}\pi^+$ (with $D^{*-} \to D^0\pi^-$)
  - $B^0 \to D^{*-}K^+$ (with $D^{*-} \to D^0\pi^-$)
  - $B^0 \to D^0\rho^0$ (with $\rho^0 \to \pi^+\pi^-$)
  - $B^0 \to D^0K^{*0}$ (with $K^{*0} \to K^+\pi^-$)

Modes with one missing kaon are not considered here as the mass shift due to the kaon missing invariant mass is too large and gives contributions only in the region $<5000\,\mathrm{MeV/c^2}$. In these decays, the mass spectrum is entirely defined by the masses of the intermediate particles and their spins (there is only one helicity amplitude in each decay).

Decays with D* Resonances

The decay chain is formed by subsequent two-body decays and the kinematic is thus fully constrained by the angular distributions of the decay products. In the reconstruction of $B \to [hh]_D h$ candidates, the D^0 meson and the bachelor track are correctly identified, although misidentification is still possible on the bachelor track, while the observed missing invariant mass is due to the non reconstructed soft pion/photon coming from the D^* meson. The shift from the nominal B mass depends on the momentum carried by this lost particle which, in turn, depends on its relative angle with respect to the bachelor track in the B rest frame. In order to study

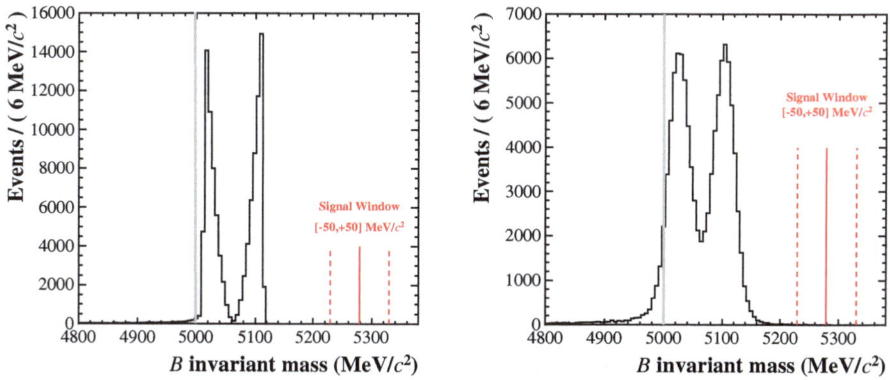

Fig. 4.8 $B^{\pm} \to D^{*0}\pi^{\pm}$ simulation (with $D^{*0} \to D^{0}\pi^{0}$) One missing *Left* true 4-momentum combination only. *Right* full detector simulation. The chosen decay model is in agreement with the correct spin assignment of the intermediate resonances: Vector $\to$ Scalar-Scalar for $D^{*0} \to D^{0}\pi^{0}$ $(1^{-} \to 0^{-}0^{-})$

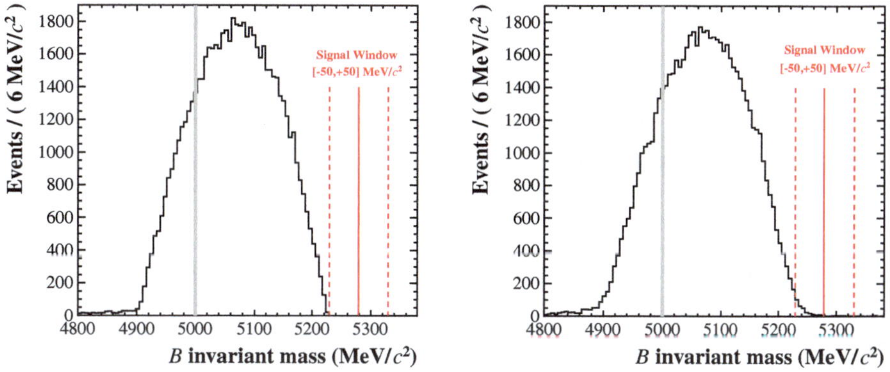

Fig. 4.9 $B^{\pm} \to D^{*0}\pi^{\pm}$ simulation (with $D^{*0} \to D^{0}\gamma$). *Left* true 4-momentum combination only. *Right* full detector simulation. The chosen decay model is in agreement with the correct spin assignment of the intermediate resonances: Vector $\to$ Scalar-Vector for $D^{*0} \to D^{0}\gamma$ $(1^{-} \to 0^{-}1^{-})$

these dependencies, each decay has been simulated exclusively with the correct spin assignment for each particle, with and without taking into account detector effects like the material interaction. Examples of B mass distributions in this kind of decays are shown in Figs. 4.8 and 4.9.

Decays with ρ^0 or K^{*0} Resonances

When the background involves a broader resonance, like ρ^0 or K^{*0}, the D^0 meson is produced directly from the B meson and one of the daughter particles of the ρ^0 or K^{*0} is lost in reconstruction. A similar argument about two-body decays applies

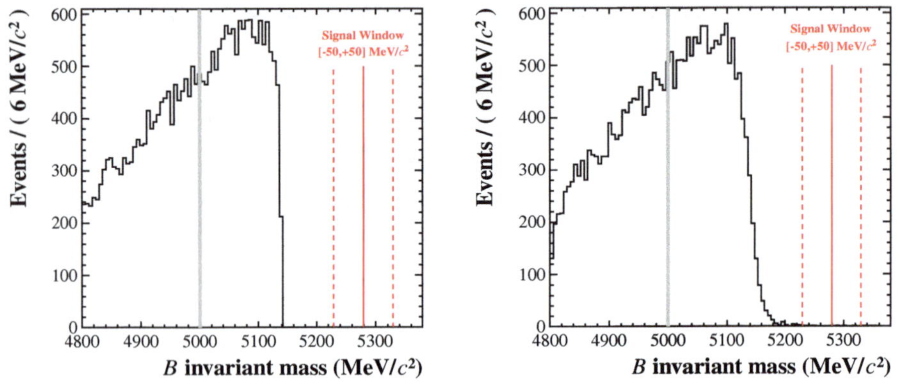

Fig. 4.10 $B^0 \to D^0\rho^0$ simulation (with $\rho^0 \to \pi^+\pi^-$). *Left* true 4-momentum combination only. *Right* full detector simulation. The mass distribution extends to lower values due to the broad phase space available in the ρ^0 meson decay

here, although more momentum configurations are available to the missing particle in the decay of the heavy resonances. Thus, the B mass distributions appear to be rather flat with a kinematic edge approaching the nominal B mass. The distribution extends to lower mass values due to the broad phase space available for the decay products, as shown in Fig. 4.10.

4.3.2 Partially Reconstructed Inclusive Sample and PDFs

After an individual detailed description of the most important expected sources of partially reconstructed backgrounds, one can look at more inclusive hadronic $B \to DX$ simulated samples to include other possible minor contributions neglected in Sect. 4.3.1. This is the case of multi-body modes like $B \to [hh]_D hhh$ decays, where more than one particle is lost in reconstruction, although their impact is expected to be significant only in the region below 5050 MeV/c^2. Other sources of background can also rise from partially reconstructed decays involving B_s and Λ_b hadronic decays, due to their heavier mass with respect to that of the $B^{\pm}$ meson. This can compensate any missing invariant mass deriving from non reconstructed particles in the decay chain. For this reason, four very large inclusive MC samples of *beauty to open charm* decays have been centrally produced by the LHCb collaboration. These decays involve the production of B_u B_d B_s mesons and Λ_b baryons decaying into charmed particles and hadrons:

- $B_u \to DX$: 40M events generated (hadronization factor 0.40)
- $B_d \to DX$: 40M events generated (hadronization factor 0.40)
- $B_s \to DX$: 10M events generated (hadronization factor 0.10)
- $\Lambda_b \to DX$: 10M events generated (hadronization factor 0.10)

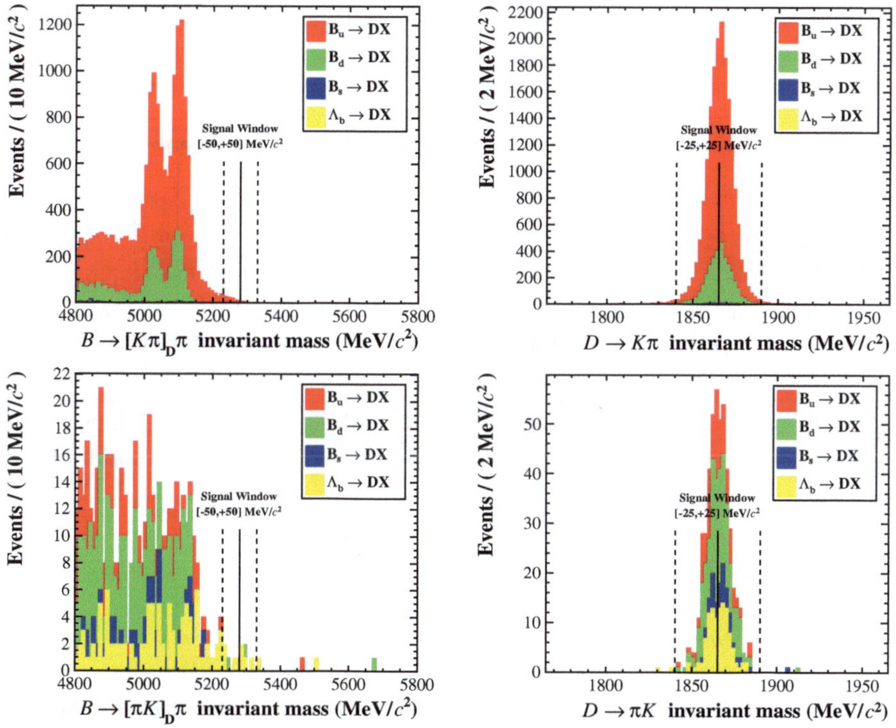

Fig. 4.11 $B \to DX$ cocktail reconstructed as $B \to [K\pi]_D\pi$ favoured mode (*Top*) and as $B \to [\pi K]_D\pi$ suppressed mode (*Bottom*). No PID requirement is applied for the bachelor and the D meson daughter tracks. The B mass distribution is plotted selecting only the events inside the D mass signal window

This inclusive hadronic cocktail has been generated in the proportions expected from the $b\bar{b}$ hadronisation factors measured experimentally in the b quark fragmentation processes [3, 4]. All the events have been reconstructed under all the $B \to [hh]_D h$ mass hypotheses and selected applying the multivariate selection presented in Sect. 3.5, excluding truth matched genuine signal $B \to [hh]_D h$ events which are treated separately in the signal components. The partially reconstructed background distributions are shown in Figs. 4.11 and 4.12 for the four different $[K\pi, \pi K]_D$ and $[KK, \pi\pi]_D$ hypotheses.

The plots for the favored $[K\pi]_D$ mode show the expected behaviour in the B and D mass distributions; the only contributions come from true D^0 mesons correctly reconstructed in B_u and B_d decays, while no sizable background from the B_s and Λ_b samples is found. Looking at the $[KK, \pi\pi]$ *CP* hypotheses, the same general pattern is found for the B_u and B_d subsamples (although with less statistics due to the lower relative branching fractions) derived from correctly reconstructed $D \to KK$ and $D \to \pi\pi$ events, with the important non-negligible contribution of Λ_b decays for the KK mass hypothesis only. For the suppressed $[\pi K]$ mode, the small number

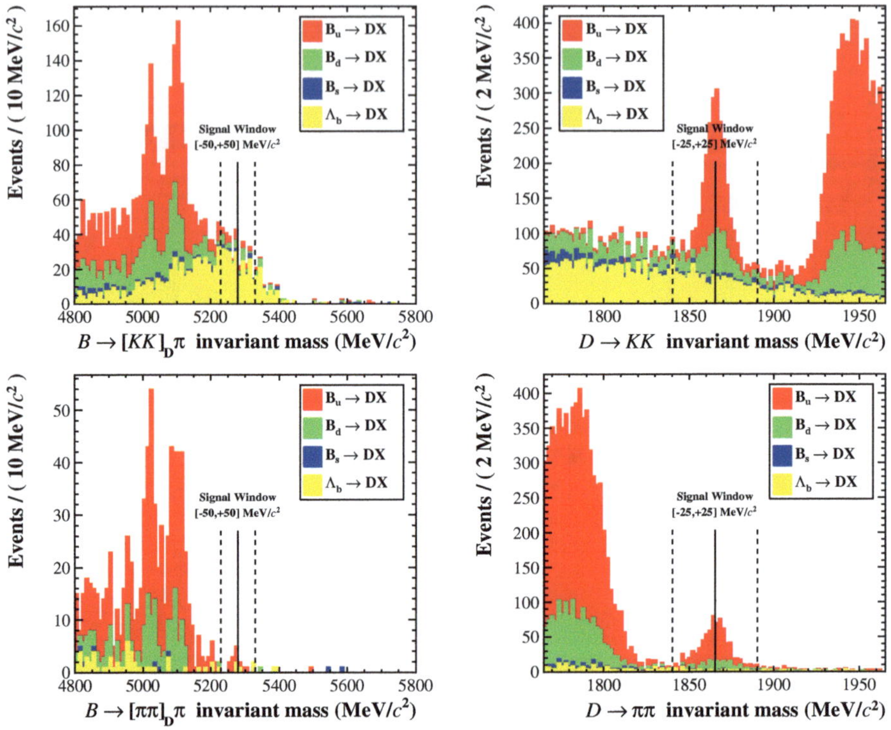

Fig. 4.12 $B \to DX$ cocktail reconstructed as $B \to [KK]_D\pi$ mode (*Top*) and as $B \to [\pi\pi]_D\pi$ mode (*Bottom*). No PID requirement is applied for the bachelor and the D meson daughter tracks. The B mass distribution is plotted selecting only the events inside the D mass signal window

of events left after the selection (even less after PID cuts are applied on the candidate tracks) is too small to shape any satisfactory PDF of the background for this particular case. The analysis thus adopts the following assumption: the *principal* part of the partially reconstructed backgrounds is the same for every mass hypotheses since it is expected to be dominated by $B \to D^*h$ events in the region >5000 MeV/c^2 with a true D^0 correctly reconstructed. This PDF, shaped and defined with the high-statistics favored mode hypothesis, is used for all the different $B \to [hh]_D h$ slices, with the two important additions described at the end of this section.
Furthermore, several improvements have been made to increase the MC/data agreement before generating the PDFs to be used in the fit:

- The simulated MC events have been rescaled as described in Sect. 3.10.
- $D\pi + X$ and $DK + X$ samples are separated using MC truth information on the bachelor track. The purpose of this procedure is to correctly reproduce the misidentification process on the bachelor tracks in partially reconstructed events, as found in data. Samples, once split according to the true mass hypothesis of the bachelor track, are weighted and recombined using the data-calibrated PID

efficiencies (see Sect. 3.11) instead of using the simulated PID response in Monte Carlo generated events.

- The semileptonic component presented in Sect. 4.2.2 is absorbed in the same PDF as the partially reconstructed hadronic events. Simulated semileptonic events have been accordingly weighted to reproduce the relative proportion found in data.
- A gaussian kernel estimation PDF is used to describe the total contribution of partially reconstructed backgrounds [5]. The widths of each gaussian are optimized to promote smoothness while ensuring a good level of detail in the low populated regions (the maximum allowed width of the gaussian kernels is set to be 25 MeV/c^2).

A complete PDF library has been generated according to different PID cuts on the bachelor track and its mass hypothesis. More details about the use of these PDFs can be found in Chap. 5, while Fig. 4.13 shows as an example the generated PDFs for a given $\Delta LL_{K/\pi}$ cut on the bachelor track of 4.

$\Lambda_b \to [Kp\pi]_{\Lambda_c}h$ for the $[KK]_D$ Slice

In the *CP* $[KK, \pi\pi]_D$ modes, the same core PDF for partially reconstructed background is used, representing decays with a true D^0 meson, with the only exception being that Λ_b decays need special consideration in $[KK]_D$ as shown in Fig. 4.12. The inclusive $B \to DX$ simulated sample, reconstructed both under the $B \to [KK]_D\pi$ and $B \to [KK]_DK$ hypotheses, is shown in Fig. 4.14.
Looking at the MC truth of the simulated events, it is found that all the Λ_b contribution is formed by $\Lambda_b \to [Kp\pi]_{\Lambda_c}\pi$, where the Λ_c daughter kaon and proton are reconstructed as the signal D^0 daughter kaons and the pion is lost. The higher mass of the Λ_c (2286.46 ± 0.14 MeV/c^2 [1]) brings $[Kp\pi]_{\Lambda_c}$ events towards to the D^0 mass signal region, even if the pion is lost in reconstruction, and the distribution

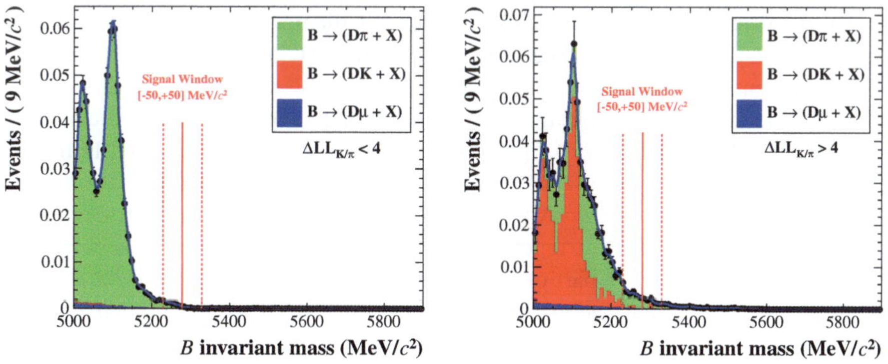

Fig. 4.13 Partially reconstructed PDFs for a $\Delta LL_{K/\pi}$ cut of 4 on the bachelor track (as used in the final simultaneous fit to data). *Left* $B \to D\pi$ hypothesis (with $\Delta LL_{K/\pi} < 4$), *Right* $B \to DK$ hypothesis (with $\Delta LL_{K/\pi} > 4$)

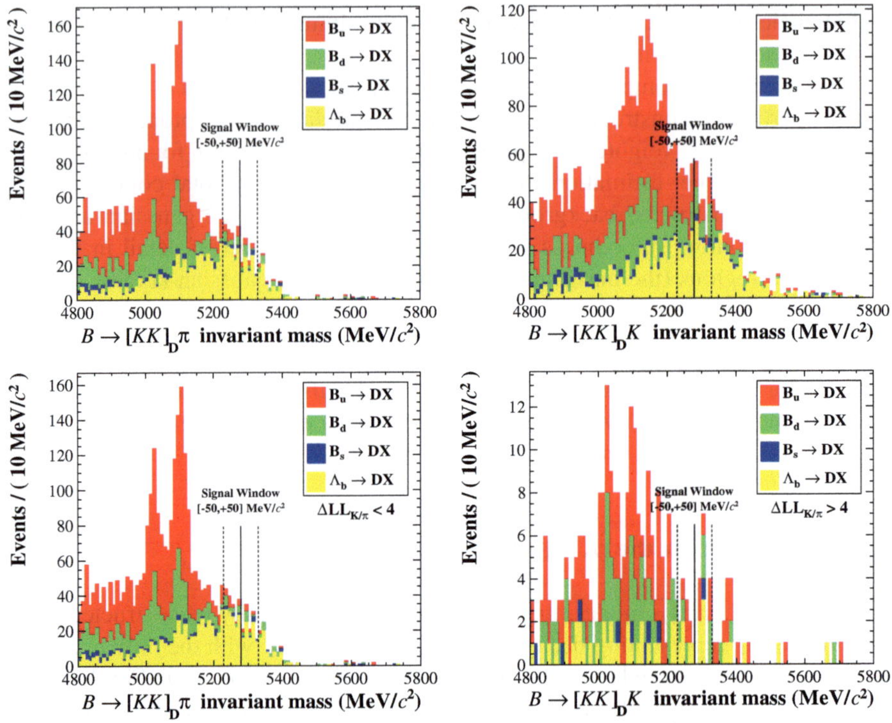

Fig. 4.14 Partially reconstructed sample reconstructed as $B \to [KK]_D\pi$ (*Left*) and $B \to [KK]_D K$ (*Right*). The *top plots* are created without PID requirements while the *bottom plots* include the specified $\Delta LL_{K/\pi} = 4$ cuts. The contribution of Λ_b decays is shown in *yellow* and extends above the nominal B mass window

in the D invariant mass for those events is expected to be approximately flat in the region considered as shown in Fig. 4.15. An additional PDF to be included in the fit, as described in Chap. 5, is thus shaped using a dedicated exclusive MC sample of the $\Lambda_b \to [Kp\pi]_{\Lambda_c}\pi$ mode generated with the full LHCb simulation. The obtained PDF is presented in Fig. 4.16 for both $B \to D\pi$ and $B \to DK$ hypotheses.
This approach is chosen, rather than simply modeling a separate, composite partially reconstructed PDF for the $B \to [KK]_D h$ fit, because of the measured branching fraction of this Λ_b decay $\mathcal{B} = (4.4 \pm 2.0) \times 10^{-4}$ [1] and the Λ_b hadronisation factor. As this mode suffers a 45 % experimental uncertainty, excluding further relative uncertainties deriving from trigger and reconstruction, it is better to separate this contribution from the *core* part of the partially reconstructed PDF and let the yield of this background float independently. This choice has been validated with toy experiments (see Sect. 5.2) to ensure an adequate fitter stability and the fitted yield of this component is in good agreement with its expected value as calculated in Table 4.1.

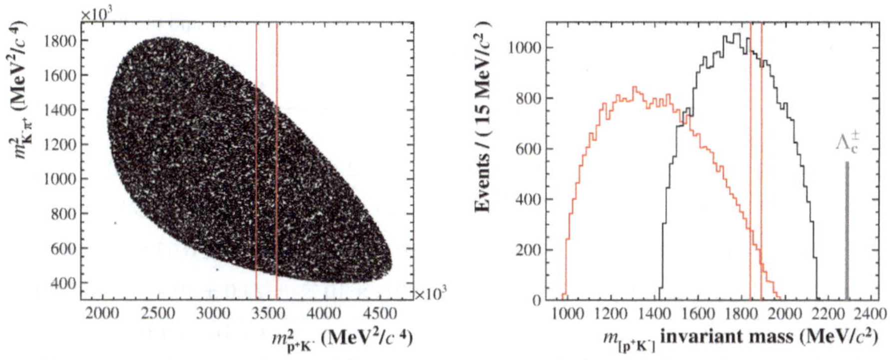

Fig. 4.15 *Left* Dalitz plot of the $\Lambda_c \to Kp\pi$ simulated decay generated with a uniform phase space distribution (true 4-momentum combination only, without any selection applied). *Right* distributions of the $[p^+K^-]$ invariant mass when the correct mass hypothesis is given on both tracks (*Black*, right distribution) and when $p \leftrightarrow K$ misidentification is supposed on the proton track to simulate the $D \to KK$ reconstruction (*Red*, left distribution). In both plots, the nominal D signal window is highlighted with *red lines*

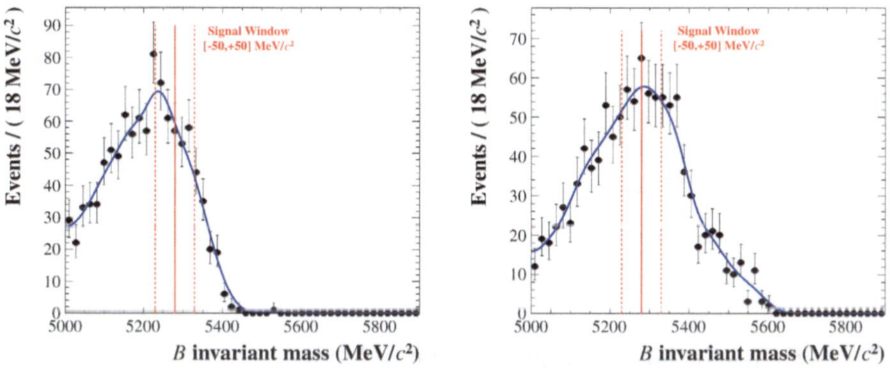

Fig. 4.16 $\Lambda_b \to [Kp\pi]_{\Lambda_c}\pi$ simulated events reconstructed as $B \to [KK]_D\pi$ (*left*) and $B \to [KK]_D K$ (*right*) with the full LHCb simulation. A gaussian kernel PDF is used in both cases. Selection is applied including the cut on the nominal D mass window and BDT

A small component for the currently unobserved[1] $\Lambda_b \to [Kp\pi]_{\Lambda_c}K$ is included as well in the fit. This is assumed to have the identical shape as its pion Cabibbo-favoured counterpart under the kaon mass hypothesis. The assigned yield is fixed as $(6.0 \pm 1.5)\,\%$ of the floated $\Lambda_b \to \Lambda_c\pi$ yield and this guess is justified when considering that non-colour-suppressed Cabibbo suppression is $\sim$7.6 % in B $\to$ Dh [1] and $\sim$4.0 % in B $\to$ ψh decays [6].

---

[1] At the time of writing, no experimental measurement is available, although it is being studied by the LHCb collaboration.

Table 4.1 Comparison between the expected yields from MC simulations and the fitted yields on data (the Λ_b contribution is normalized to the $B \to [KK]_D\pi$ mode)

	$\Lambda_b \to [Kp\pi]_{\Lambda_c}\pi$	$B \to [KK]_D\pi$	Ratio $\frac{\Lambda_b \to [Kp\pi]_{\Lambda_c}\pi}{B \to [KK]_D\pi}$
f_{had}	0.1	0.4	0.25
$\varepsilon_{\text{eff}}^{\text{Geom}}$	0.1634 ± 0.0003	0.1604 ± 0.0002	1.018 ± 0.002 (0.21 % rel.)
$\varepsilon_{\text{eff}}^{\text{Sel}}$	$(1.98 \pm 0.10) \times 10^{-4}$	$(2.79 \pm 0.02) \times 10^{-2}$	0.0071 ± 0.0004 (5.0 % rel.)
$\varepsilon_{\text{eff}}^{\text{Trig}}$	0.95 ± 0.07	0.96 ± 0.01	0.99 ± 0.07 (7.5 % rel.)
$\mathcal{B}$	$(4.40 \pm 1.97) \times 10^{-4}$	$(1.92 \pm 0.07) \times 10^{-5}$	23.0 ± 10.3 (44.9 % rel.)
Expected			$\mathbf{0.041 \pm 0.019}$(45.8 % rel.)
Yield in data	808 ± 60	13343 ± 187	$\mathbf{0.061 \pm 0.005}$(7.6 % rel.)

f_{had} is the expected hadronisation factor [4], while $\varepsilon_{\text{eff}}^{\text{Geom}}$, $\varepsilon_{\text{eff}}^{\text{Sel}}$ and $\varepsilon_{\text{eff}}^{\text{Trig}}$ are the geometrical, selection and trigger efficiencies respectively obtained from MC simulated events as introduced in Sect. 3.8. Branching Fractions are taken from [1], the main source of uncertainty being the branching fraction of $\Lambda_b \to \Lambda_c\pi$. The expected ratio $\mathbf{n}_{\Lambda_b \to \Lambda_c\pi}/\mathbf{n}_{B \to [KK]_D\pi}$ is compatible at the level of $\sim 1\sigma$ with respect to what is found in data (last row of the table)

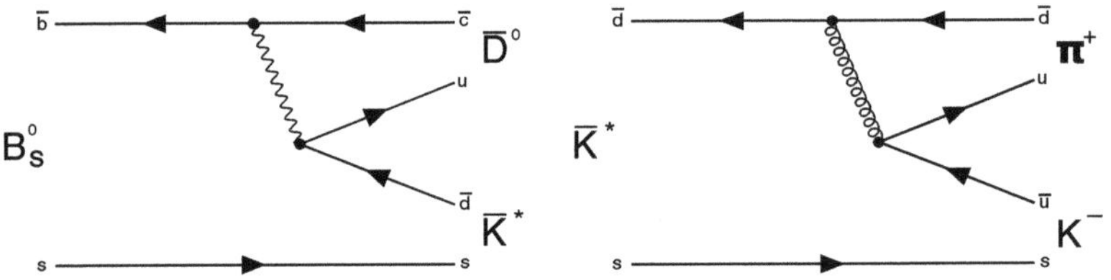

Fig. 4.17 Feynman diagrams for the $B_s \to [K\pi]_{D^0}[K\pi]_{K^*}$ decay. The diagram for the favoured $\bar{D}^0 \to K^+\pi^-$ decay is already presented in Fig. 1.6

$B_s \to [K\pi]_{D^0}K\pi$ for the Suppressed $[\pi K]_D$ Mode

B_s^0 decays, which are infrequent compared to the favoured $B_{u,d}$ decays, due to $f_s/f_d \sim 0.25$, cannot be neglected for the suppressed $B \to [\pi K]_D h$ mode. The most important contribution to be considered is $B_s \to D^0[K\pi]K\pi$ with the D^0 meson correctly reconstructed as $[\pi K]_D$ and the bachelor pion is lost. This mode involves a colour-suppressed but Cabibbo-allowed $b \to c$ transition as shown in Fig. 4.17 and contains a real D meson along with opposite-sign kaons in the final state. This experimental signature is identical to that of the suppressed $B \to [\pi K]_D K$ signal mode and the higher mass of the B_s meson brings the kinematic edge of the background distribution closer to the nominal B signal window. It can be also noted that any equivalent $B_s \to D^0\pi\pi$ background is Cabibbo-suppressed. $B_s \to [K\pi]_{D^0}[K\pi]_{K^*}$ resonant events have been generated with the full LHCb detector simulation and reconstructed as $B \to [\pi K]_D K$.

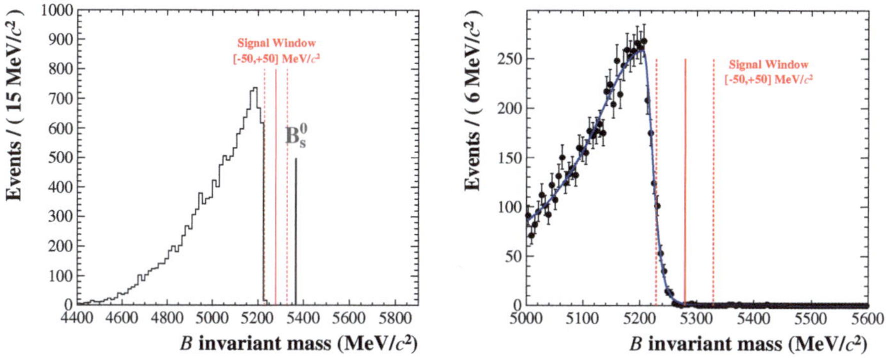

Fig. 4.18 $B_s^0 \to D^0[K\pi]_{K^*}$ events. *Left* true 4-momentum combination only in the full mass window when the bachelor pion is lost. *Right* PDF shaped using the full LHCb detector simulation. The distribution is fitted using a bifurcated gaussian with two tail parameters (Sect. 3.9.1)

Table 4.2 Comparison between the expected yields from MC simulations and the fitted yields on data with the $B_s \to [K\pi]_{D^0}[K\pi]_{K^*}$ contribution normalized to the favourite $B \to [K\pi]_D\pi$

	$B_s \to [K\pi]_{D^0}[K\pi]_{K^*}$	$B \to [K\pi]_D\pi$	Ratio $\frac{B_s \to [K\pi]_{D^0} K^*}{B \to [K\pi]_D\pi}$
f_{had}	0.1	0.4	0.25 ± 0.03 (12.4 % rel.)
$\varepsilon_{\text{eff}}^{\text{Geom}}$	0.1569 ± 0.0004	0.1604 ± 0.0002	0.978 ± 0.003 (0.28 % rel.)
$\varepsilon_{\text{eff}}^{\text{Sel}}$	$(2.27 \pm 0.30) \times 10^{-3}$	$(2.55 \pm 0.01) \times 10^{-2}$	0.09 ± 0.01 (13.0 % rel.)
$\varepsilon_{\text{eff}}^{\text{Trig}}$	0.95 ± 0.07	0.95 ± 0.01	1.00 ± 0.07 (7.4 % rel.)
$\mathcal{B}$	$2 \times (1.22 \pm 0.36) \times 10^{-5}$	$(1.87 \pm 0.06) \times 10^{-4}$	0.130 ± 0.039(29.6 % rel.)
Expected			$\mathbf{0.0027 \pm 0.0010}$(35.4 % rel.)
Yield in data	-	-	$\mathbf{0.0029 \pm 0.0004}$(14.0 % rel.)

f_{had} is the measured hadronisation factor [4], while $\varepsilon_{\text{eff}}^{\text{Geom}}$, $\varepsilon_{\text{eff}}^{\text{Sel}}$ and $\varepsilon_{\text{eff}}^{\text{Trig}}$ are the geometrical, selection and trigger efficiencies respectively obtained from MC simulated events as introduced in Sect. 3.8. The resonant $B_s \to DK^*$ branching fraction is taken from [7] and about half of the bachelor $K\pi$ system is supposed to be in the K^* resonance [1]. The error on the branching fraction is the main source of uncertainty in the estimation (about 30 % relative). The expected ratio $\mathbf{n_{B_s \to DK^*}/n_{B \to [K\pi]_D\pi}}$ is compatible with the measured one

A dedicated PDF has been thus shaped on the offline selected events and used in the signal extraction fit as a fixed component with floating yield (relative to the favoured $B \to [K\pi]_D\pi$ mode). The $B_s \to [K\pi]_{D^0}[K\pi]_{K^*}$ PDF is shown in Fig. 4.18. Simulated events are also used to estimate the expected $B_s \to D^0 K\pi$ yield in data as presented in Table 4.2. A systematic uncertainty is then assigned on the shape of this component obtained varying the mean and widths of the PDF on data as described in Sect. 5.3).

Table 4.3 Charmless backgrounds for the $B \to [hh]_D h$ signal modes

Charmless mode		Simulated	Branching ratio [1]
$B^+ \to \pi^+\pi^-K^+$	(RES)	Yes	$(5.10 \pm 0.29) \times 10^{-5}$
	(PHSP)	Yes	$\left(1.63^{+0.21}_{-0.15}\right) \times 10^{-5}$
$B^+ \to K^+K^-K^+$	(RES)	No	$(3.37 \pm 0.22) \times 10^{-5}$
	(PHSP)	Yes	$\left(2.8^{+0.9}_{-1.6}\right) \times 10^{-5}$
$B^+ \to \pi^+\pi^-\pi^+$	(RES)	No	$(1.52 \pm 0.14) \times 10^{-5}$
	(PHSP)	Yes	$\left(5.3^{+1.5}_{1.1}\right) \times 10^{-6}$
$B^+ \to K^+K^-\pi^+$	(RES)	Yes	$(5.0 \pm 0.7) \times 10^{-6}$
	(PHSP)	Yes	$<7.5 \times 10^{-5}$ (CL = 90 %)
$B^+ \to K^+\pi^-K^+$	(RES)	No	$<1.6 \times 10^{-7}$ (CL = 90 %)
$B^+ \to \pi^+K^-\pi^+$	(RES)	No	$<9.5 \times 10^{-7}$ (CL = 90 %)

For each mode the table reports both the total measured resonant branching fraction (RES = Resonant) and its correspondent non-resonant part (PHSP = Phase Space, uniform distribution on the Dalitz plot). The measured $\mathcal{B}$ values are known with very high uncertainty (up to a relative 50 % in most cases), thus this uncertainty unavoidably reflects on any possible estimation of the charmless yields in data from MC. The third column indicates whether the mode has been simulated exclusively or not

4.4 Charmless Backgrounds

The purpose of this section is to investigate any possible contribution from charmless backgrounds. Charmless modes have the same $B \to hhh$ final state of $B \to [hh]_D h$ decays, but they don't proceed through the $D^0 \to hh$ intermediate state. In this study, every charge combination of the three final state hadrons is considered with both kaon and pion mass hypotheses. Charmless modes are expected to peak at the nominal B mass with a width comparable to that of the signal, while being relatively flat in the D mass distribution (depending on the particular Dalitz plot configuration). They can thus give potential biases on the branching fraction ratio extraction and even interfere with the CP asymmetry of the $B \to [hh]_D h$ modes. All the $B \to hhh$ modes considered are shown in Table 4.3 along with their measured branching fractions taken from [1].

In principle, charmless events can have reconstruction efficiencies comparable those found for $B \to [hh]_D h$ signal events. However, any cut on the D^0 invariant mass strongly reduces their total effective reconstruction efficiency, selecting only a tiny portion of the Dalitz plot. An example of this effect is shown in Fig. 4.19 for $B^+ \to \pi^+\pi^-K^+$ events with two different model assumptions. The analysis focuses only on those modes that can give a sizable contribution compared to the expected number of $B \to [hh]_D h$ signal events in the offline selected sample. This is particularly true for the $[\pi K]_D$ suppressed mode, due to its small expected branching fraction of $\mathcal{O}(10^{-7})$, and the $[KK, \pi\pi]_D$ CP modes, whose branching ratios are comparable to their respective charmless modes, once efficiency suppressions deriving from the D mass cut are taken into account.

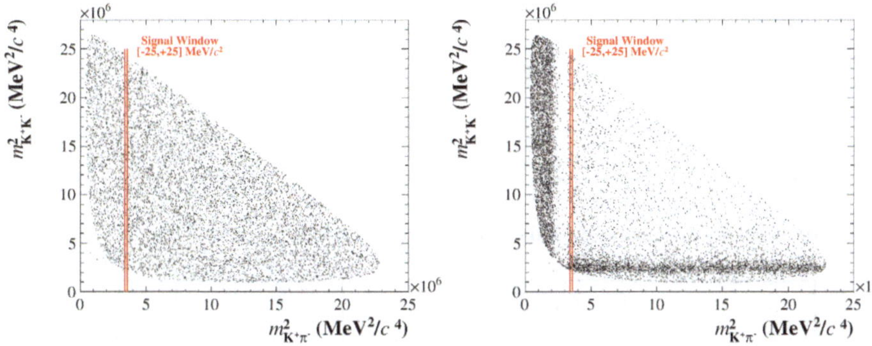

Fig. 4.19 Dalitz plots for $B^+ \rightarrow \pi^+\pi^-K^+$ events with two different model assumptions. Both plots are generated using true 4-momentum combination only. *Left* uniform phase space (PHSP) non-resonant model; in this configuration, the chosen cut on the D^0 signal mass window $[-25, +25]$ MeV/c^2 selects about (1.1 ± 0.2) % of the generated events in the full Dalitz space. *Right* simulated $B^+ \rightarrow \pi^+\pi^-K^+$ resonant model (RES). This model is taken from [8] and it is chosen to approximately match the Dalitz plot distribution seen on data as presented in that paper. Under this assumption, (1.3 ± 0.2) % of the generated events falls inside the D^0 signal window

The impact of charmless events in the final sample is further reduced by introducing an additional cut on the flight distance of the D^0 meson, as described in Sect. 4.4.1. The remaining charmless contribution is carefully double-checked respectively using both expectations from MC simulated events in Sect. 4.4.2 and D^0 data sidebands, where available, in Sect. 4.4.3. In all the sections, only charmless modes that give a sizeable contribution in the final sample will be considered (e.g. >1 expected event after the full selection is applied).

4.4.1 Selecting on D Flight Distance

In order to reduce charmless backgrounds the analysis exploits the high forward boost of genuine signal candidates in the LHCb detector along the longitudinal direction, parallel to the beam axis. A specific cut is introduced on the signed flight distance significance of the reconstructed D meson with respect to the B meson vertex in the z-direction, defined as (Fig. 4.20):

$$FD_{[z_D^{vtx} - z_B^{vtx}]} = \frac{z_D^{vtx} - z_B^{vtx}}{\sqrt{\sigma^2_{z_D} + \sigma^2_{z_B}}}$$

In fact, $B \rightarrow [hh]_D h$ signal events are distributed mostly in the *forward* region with just a few percent of the true D candidates found upstream of the B meson vertex. In contrast, charmless decays are expected to be approximately symmetric around zero (the exact distribution depends slightly on the particular 3-body decay

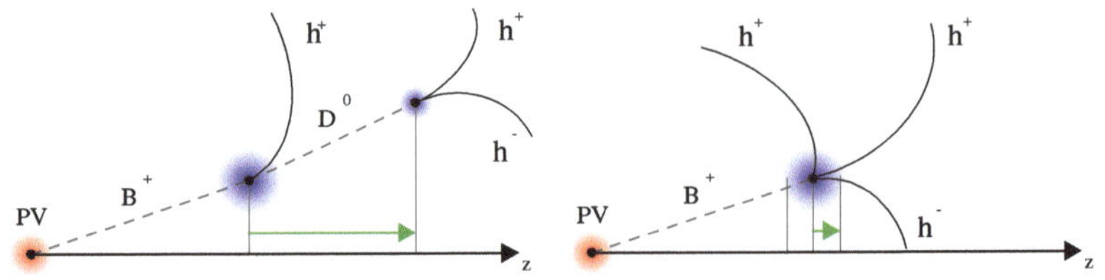

Fig. 4.20 Topology of signal (*left*) and charmless (*right*) decays

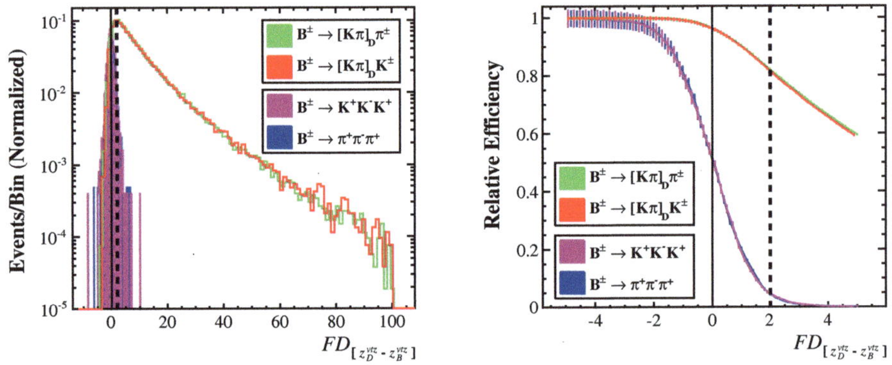

Fig. 4.21 *Left* $FD_{[z_D^{vtx}-z_B^{vtx}]}$ distributions for signal and charmless MC simulated events. Distributions have been normalized and plotted on a logarithmic scale in order to highlight the long exponential-like tail in genuine $B \to [hh]_D h$ decays. *Right* corresponding efficiencies for each $FD_{[z_D^{vtx}-z_B^{vtx}]}$ cut

model), with a modest bias towards the positive part of the z-axis due to the LHCb acceptance as illustrated in Fig. 4.21, Tables 4.4 and 4.5. B candidates are required to have $FD_{[z_D^{vtx}-z_B^{vtx}]} > 2$; this cut has been chosen in order to reduce the expected level of charmless background in the final sample to a negligible amount (with the expected $n_{hhh}^{expect} \ll \sigma_{B\to[hh]_D h}^{stat}$ the statistical uncertainty on the signal yield), while ensuring an efficiency on true signal candidates of about 82 %. The agreement of MC simulation and data in the $FD_{[z_D^{vtx}-z_B^{vtx}]}$ variable is also double-checked fitting the most abundant $B \to [K\pi]_D\pi$ control sample at different $FD_{[z_D^{vtx}-z_B^{vtx}]}$ cuts, with a procedure similar to the one described in Sect. 3.5.2, and comparing the efficiencies found on data to the one obtained with MC simulated events. This procedure, presented in Fig. 4.22, suggests an excellent MC/data agreement on the full range of the variable.

4.4.2 Charmless Expectations from MC

Four charmless modes have been generated using the full LHCb detector simulation, assuming a non resonant flat Dalitz distribution. Each sample has been reconstructed

Table 4.4 Efficiencies of $B \to [K\pi]_D h$ MC simulated samples for different $FD_{[z_D^{vtx}-z_B^{vtx}]}$ cuts (in percent %)

$FD_{[z_D^{vtx}-z_B^{vtx}]}$	$B \to [K\pi]_D\pi$	$B \to [K\pi]_D K$
>0	96.3 ± 0.6	96.3 ± 0.6
>2	82.2 ± 0.5	81.6 ± 0.5
>4	66.5 ± 0.5	65.6 ± 0.5

Errors are statistical only and depend on the number of generated and selected MC events for each mode

Table 4.5 Efficiencies of charmless $B \to hhh$ MC simulated samples for different $FD_{[z_D^{vtx}-z_B^{vtx}]}$ cuts (in percent %)

$FD_{[z_D^{vtx}-z_B^{vtx}]}$	$B^+ \to K^+K^-K^+$	$B^+ \to \pi^+\pi^-\pi^+$	$B^+ \to \pi^+K^-K^+$	$B^+ \to \pi^+\pi^-K^+$
>0	51.6 ± 1.9	51.1 ± 1.8	48.98 ± 8.6	48.8 ± 2.6
>2	4.7 ± 0.5	4.5 ± 0.43	5.10 ± 2.3	3.7 ± 0.5
>4	0.14 ± 0.08	0.18 ± 0.13	–	0.19 ± 0.13

Errors are statistical only and depend on the number of generated and selected MC events for each mode

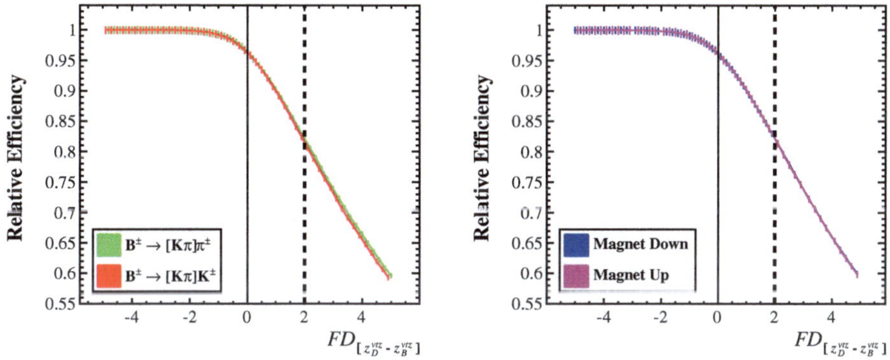

Fig. 4.22 *Left* efficiencies for MC simulated $B \to [K\pi]_D h$ decays. *Right* efficiencies for the fitted $B \to [K\pi]_D\pi$ control sample on data with different magnet polarizations. Excellent agreement is found between simulation and data

under every $B \to [hh]_D h$ mass hypothesis and the most important contributions have been recognized for each $[hh]_D$ slice. Both the B and D mass distributions have been investigated before and after applying the constraint on the D mass described in Sect. 3.9.2 In each case, the D mass distributions appear to be rather flat in the wide D mass window of $[-100, +100]$ MeV/c^2 (as used in the preliminary reconstruction of the D candidates) around the nominal central value [1]. The $[hh]_D$ mass distributions for the largest available charmless samples are shown in Fig. 4.23. Assuming a completely uniform distribution, a relative efficiency of about 25 % is thus expected when selecting candidates inside the nominal signal window of $[-25, +25]$ MeV/c^2 from its central value. B mass distributions, instead, are expected

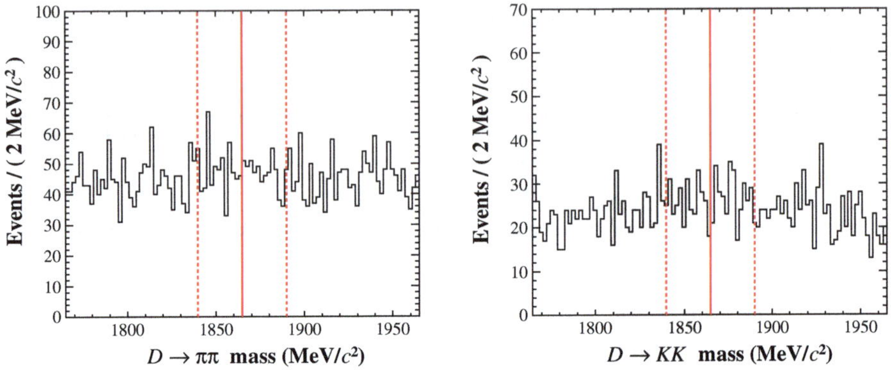

Fig. 4.23 D mass distributions for charmless MC simulated events (preliminary selection only). *Left* $[\pi\pi]_D$ from $B \to \pi^+\pi^-\pi^+$ events. *Right* $[KK]_D$ from $B \to K^+K^-K^+$ events. The D signal window is highlighted with *vertical bars*

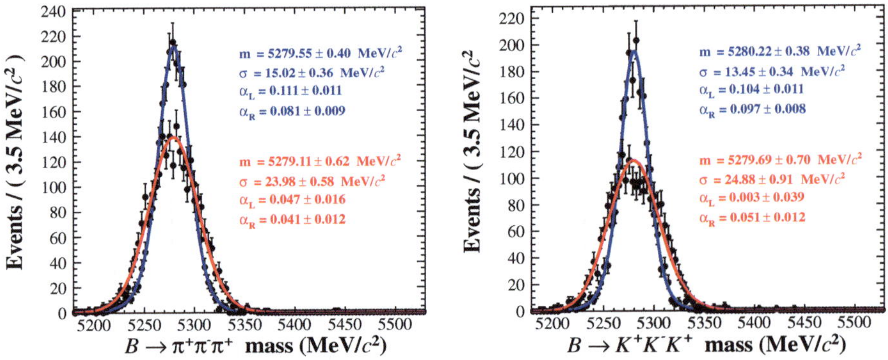

Fig. 4.24 B mass distributions for charmless MC simulated events ($m_D \in [-25, +25]$ MeV/c^2 applied). *Blue* (*narrow curve*): before D mass constraint, Red (*broad curve*): after D mass constraint

to be peaking at the nominal B mass with widths comparable to those of signal events (the widths depend on the chosen D mass window). The effect of the D mass constraint on the B mass distribution is then assessed in Fig. 4.24. While reducing the width of the genuine signal peaks as described in Sect. 3.9.2, it increases the width of charmless background peaks from about 14 MeV/c^2 to about 24 MeV/c^2, providing additional benefits with the dilution of these background events over a wider mass range. D-mass constrained charmless peaks should be compared to the typical widths of signal $B \to [hh]_D h$ MC simulated events (about 11.5 MeV/c^2 as shown in Fig. 3.16), therefore they are expected to be twice as broad as the signal.

Using the simulated samples, an estimation of the expected remaining charmless yields in the offline selected sample can be provided. Every charmless $B \to hhh$ yield has been compared to its relevant $B \to [hh]_D h$ mode, e.g. the signal mode of which it can be a source of background. Note that different modes can be recon-

Table 4.6 Expectations of charmless background yields for $B \to [KK]_D h$ signal mode

	$B \to K^+K^-K^+$ **reco as** $B \to [KK]_D K$	$B \to \pi^+K^-K^+$ **reco as** $B \to [KK]_D \pi$
Charmless MC	1660492	157100
Selection	6488	276
$m_D \in [-25, +25]$ MeV/c^2	1684	79
$FD_{[z_D^{vtx} - z_B^{vtx}]} > 2$	87	5
Charmless $B \to hhh$ $\varepsilon_{\text{eff}}^{\text{Sel}}$	$(5.24 \pm 0.56) \times 10^{-5}$	$(3.18 \pm 1.42) \times 10^{-5}$
Charmless $B \to hhh$ $\mathcal{B}$	$(3.37 \pm 0.22) \times 10^{-5}$	$(5.00 \pm 0.70) \times 10^{-6}$
Charmless $B \to hhh$ $\varepsilon_{\text{eff}}^{\text{Geom}}$	0.1698 ± 0.0005	0.1649 ± 0.0005
Total	$(3.00 \pm 0.38) \times 10^{-10}$	$(2.62 \pm 1.23) \times 10^{-11}$
Signal $B \to [KK]_D h$ $\varepsilon_{\text{eff}}^{\text{Sel}}$	$(3.23 \pm 0.02) \times 10^{-2}$	$(3.26 \pm 0.02) \times 10^{-2}$
Signal $B \to [KK]_D h$ $\mathcal{B}$	$(1.46 \pm 0.13) \times 10^{-6}$	$(1.92 \pm 0.07) \times 10^{-5}$
Signal $B \to [KK]_D h$ $\varepsilon_{\text{eff}}^{\text{Geom}}$	0.1714 ± 0.0002	0.1667 ± 0.0004
Total	$(8.07 \pm 0.74) \times 10^{-9}$	$(1.04 \pm 0.04) \times 10^{-7}$
Ratio $n_{B \to hhh}/n_{B \to [KK]_D h}$	0.0372 ± 0.0058	0.0003 ± 0.0001
Fitted $B \to [KK]_D h$ yield	1031 ± 50	13343 ± 187
Expected $B \to hhh$ **yield**	38 ± 6	3 ± 2

All the efficiencies are obtained from MC simulations (Uncertainties on efficiencies are statistical only)

structed under different mass hypotheses: for example, $B \to \pi^+K^-K^+$ decays can in principle form a background for both the $B \to [KK]_D\pi$ and the suppressed $B \to [\pi K]_D K$ slices. Selection efficiencies are taken from MC simulations for both signal and charmless events, while branching fractions are taken from Table 4.3. The estimated ratio $n_{B \to hhh}/n_{B \to [hh]_D h}$ is then used to calculate the expected number of charmless events in each $[hh]_D$ slice, as shown in Tables 4.6, 4.7 and 4.8. In all the cases the expected yield of charmless events is much lower than the statistical uncertainty of its relevant signal yield (and those events are distributed according to a Gaussian distribution with a width twice as broad); for this reason no specific charmless components are included in the fitter.

4.4.3 Charmless Expectations from Data Sidebands: $[KK, \pi\pi]_D$

For the *CP* modes $B \to [KK]_D h$ and $B \to [\pi\pi]_D h$, data sidebands can be used to double-check the expected level of charmless events in data. This totally data-driven approach exploits the relatively flat distribution of charmless events in the D mass distribution: this allows a trivial rescaling procedure into the signal window, since the same selection is applied, without using any assumption on $\varepsilon_{\text{eff}}^{\text{Sel}}$, $\varepsilon_{\text{eff}}^{\text{Geom}}$ and $\mathcal{B}$ (known with poor accuracy). The D mass spectra for $[KK]_D$ and $[\pi\pi]_D$ are shown in Fig. 4.25, along with the chosen sideband definition. In each histogram the contribution of misidentified $D \to K\pi$ favoured events is seen displaced from the D mass (for more details, see Sect. 4.5) for this reason, one can only use one sideband and look for peaking structures in the B mass. Note that no similar procedure is

Table 4.7 Expectations of charmless background yields for $B \to [\pi\pi]_D h$ signal mode

	$B \to \pi^+\pi^-\pi^+$ **reco as** $B \to [\pi\pi]_D\pi$	$B \to \pi^+\pi^-K^+$ **reco as** $B \to [\pi\pi]_D K$
Charmless MC	2220989	1690094
Selection	8284	3402
$m_D \in [-25, +25]$ MeV/c^2	2063	866
$FD_{[z_D^{vtx} - z_B^{vtx}]} > 2$	91	23
Charmless $B \to hhh$ $\varepsilon_{\rm eff}^{\rm Sel}$	$(4.10 \pm 0.43) \times 10^{-5}$	$(1.36 \pm 0.29) \times 10^{-5}$
Charmless $B \to hhh$ $\mathcal{B}$	$(3.37 \pm 0.22) \times 10^{-5}$	$(5.10 \pm 0.29) \times 10^{-5}$
Charmless $B \to hhh$ $\varepsilon_{\rm eff}^{\rm Geom}$	0.1557 ± 0.0005	0.1606 ± 0.0005
Total	$(2.15 \pm 0.27) \times 10^{-10}$	$(1.11 \pm 0.24) \times 10^{-10}$
Signal $B \to [\pi\pi]_D h$ $\varepsilon_{\rm eff}^{\rm Sel}$	$(3.46 \pm 0.02) \times 10^{-2}$	$(3.45 \pm 0.02) \times 10^{-2}$
Signal $B \to [\pi\pi]_D h$ $\mathcal{B}$	$(6.78 \pm 0.25) \times 10^{-6}$	$(5.15 \pm 0.47) \times 10^{-7}$
Signal $B \to [\pi\pi]_D h$ $\varepsilon_{\rm eff}^{\rm Geom}$	0.1540 ± 0.0002	0.1572 ± 0.0002
Total	$(3.61 \pm 0.13) \times 10^{-8}$	$(2.79 \pm 0.26) \times 10^{-9}$
Ratio $n_{B \to hhh}/n_{B \to [\pi\pi]_D h}$	0.0060 ± 0.0008	0.0399 ± 0.0094
Fitted $B \to [\pi\pi]_D h$ yield	3942 ± 98	317 ± 27
Expected $B \to hhh$ **yield**	23 ± 3	12 ± 3

All the efficiencies are obtained from MC simulations (Uncertainties on efficiencies are statistical only)

Table 4.8 Expectations of charmless background yields for $B \to [\pi K]_D h$ signal mode

	$B \to \pi^+\pi^+K^-$ **reco as** $B \to [\pi K]_D\pi$	$B \to \pi^+K^-K^+$ **reco as** $B \to [\pi K]_D K$
Charmless MC	1690094	157100
Selection	2125	189
$m_D \in [-25, +25]$ MeV/c^2	541	51
$FD_{[z_D^{vtx} - z_B^{vtx}]} > 2$	24	2
Charmless $B \to hhh$ $\varepsilon_{\rm eff}^{\rm Sel}$	$(1.42 \pm 0.29) \times 10^{-5}$	$(1.43 \pm 0.95) \times 10^{-5}$
Charmless $B \to hhh$ $\mathcal{B}$	$(9.50 \pm 9.50) \times 10^{-7}$	$(5.00 \pm 0.07) \times 10^{-6}$
Charmless $B \to hhh$ $\varepsilon_{\rm eff}^{\rm Geom}$	0.1606 ± 0.0005	0.1649 ± 0.0005
Total	$(2.17 \pm 2.21) \times 10^{-12}$	$(1.18 \pm 0.80) \times 10^{-11}$
Signal $B \to [K\pi]_D h$ $\varepsilon_{\rm eff}^{\rm Sel}$	$(2.59 \pm 0.01) \times 10^{-2}$	$(2.54 \pm 0.01) \times 10^{-2}$
Signal $B \to [K\pi]_D h$ $\mathcal{B}$	$(5.62 \pm 0.19) \times 10^{-7}$	$(2.00 \pm 2.00) \times 10^{-7}$
Signal $B \to [K\pi]_D h$ $\varepsilon_{\rm eff}^{\rm Geom}$	0.1604 ± 0.0002	0.1639 ± 0.0002
Total	$(2.33 \pm 0.08) \times 10^{-7}$	$(8.33 \pm 8.33) \times 10^{-10}$
Ratio $n_{B \to hhh}/n_{B \to [K\pi]_D h}$	0.00093 ± 0.00095	0.014 ± 0.017
Fitted $B \to [K\pi]_D h$ yield	334 ± 21	96 ± 13
Expected $B \to hhh$ **yield**	0 ± 1	1 ± 2

All the efficiencies are obtained from MC simulations (Uncertainties on efficiencies are statistical only)

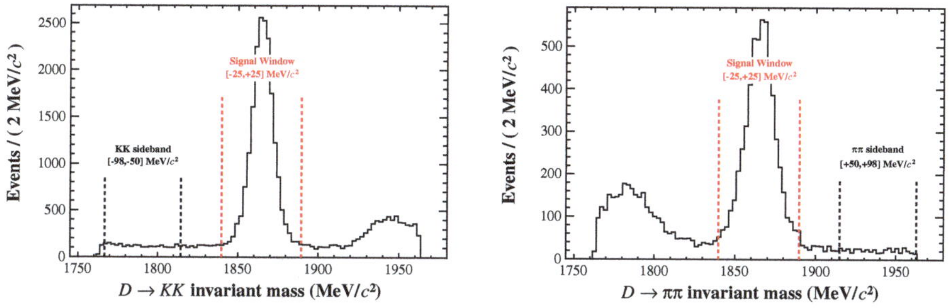

Fig. 4.25 *Left* $D \rightarrow KK$, *Right* $D \rightarrow \pi\pi$. The sideband regions are defined as $[-98, -50]$ MeV/c^2 for $[KK]_D$ and $[+50, +98]$ MeV/c^2 for $[\pi\pi]_D$ from the center of the D peak

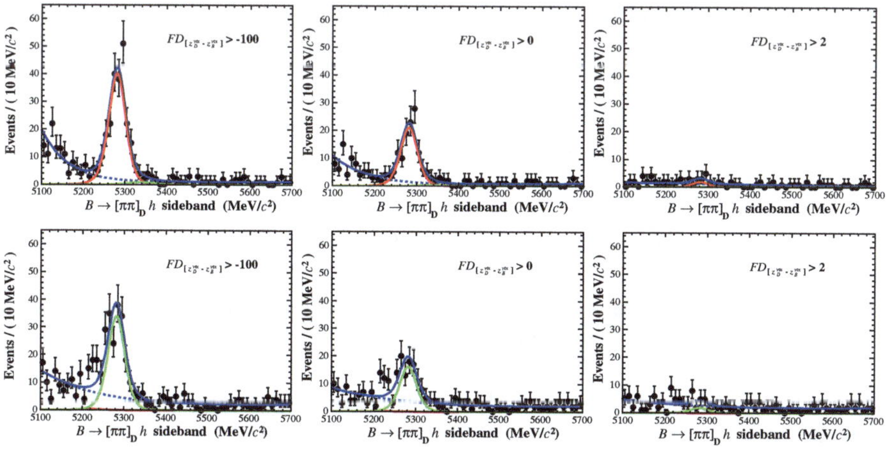

Fig. 4.26 $B^+ \rightarrow \pi^+\pi^-K^+$ (*top*) and $B^+ \rightarrow \pi^+\pi^-\pi^+$ (*bottom*) fits on the $[\pi\pi]_D$ sideband. *Red* $B \rightarrow [\pi\pi]_D K$ candidates, *Green* $B \rightarrow [\pi\pi]_D \pi$ candidates (a simultaneous fit similar to the one used in the full analysis is used)

possible for the suppressed $[\pi K]_D$ slice as the D sidebands are populated by the more abundant $B \rightarrow [KK]_D h$ (lower sideband) and $B \rightarrow [\pi\pi]_D h$ (higher sideband) modes on both sides of the D mass peak.

In order to extract the expected charmless yield in the signal region, the full selection is applied to the sidebands, including $\Delta LL_{K/\pi}$ PID cuts on the candidates' tracks. The B mass distributions of the sidebands are then fitted with a modified Gaussian shape (Sect. 3.9.1) for the signal component and an exponential function for the background. The B mass variable without any constraints on the D mass is chosen in order to avoid induced artificial distortions deriving from events with m_D far from the D^0 central value. The extracted signal-like yield is then rescaled to the nominal signal window assuming a flat distribution in the D mass region of interest. The fits on the D sidebands are shown in Figs. 4.26 and 4.27, while the expected yields for each $FD_{[z_D^{vtx}-z_B^{vtx}]}$ cut are reported in Table 4.9.

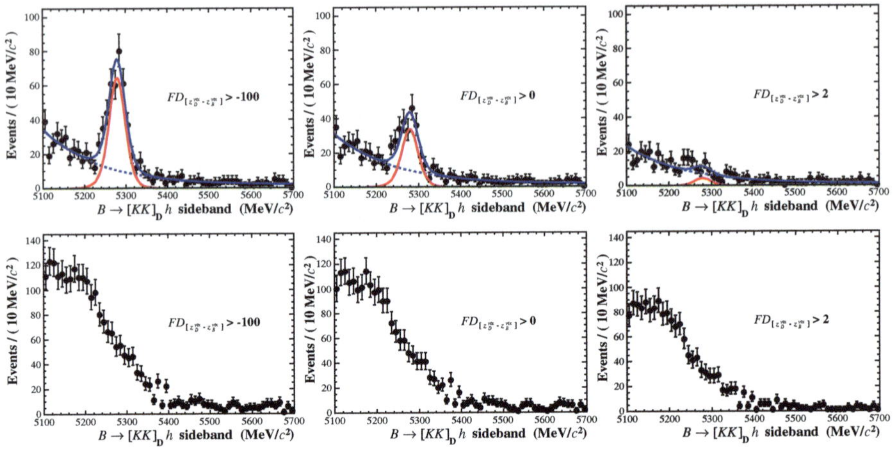

Fig. 4.27 $B^+ \rightarrow K^+K^-K^+$ (*top*) and $B^+ \rightarrow K^+K^-\pi^+$ (*bottom*) fits on the $[KK]_D$ sideband. *Red* $B \rightarrow [KK]_D K$ candidates, *Green* $B \rightarrow [KK]_D\pi$ candidates (not visible in the figures). The *bottom plots* are dominated by the $\Lambda_b \rightarrow [Kp\pi]_{\Lambda_c} h$ contribution

$B^+ \rightarrow K^+K^-\pi^+$ events could not be extracted directly from the sidebands due to the relatively small expected contribution compared to the size of the background, dominated in this case by the $\Lambda_b \rightarrow [Kp\pi]_{\Lambda_c} h$ contribution as discussed in Sect. 4.3.2. In fact, no significant excess of signal-like peaking events is found in the $[KK]_D\pi$ sideband at any $FD_{[z_D^{vtx} - z_B^{vtx}]}$ cut to give a reliable fit. Several tests have been performed to validate this data-driven procedure, e.g. checking the dependence of the charmless yields with respect to different BDT and $FD_{[z_D^{vtx} - z_B^{vtx}]}$ cuts in order to be sure that the peaking structures found in the sidebands could be interpreted as charmless events. Yields extracted on data are found to rescale exactly as expected from MC simulation, as shown in Fig. 4.28. This is also a direct demonstration of both the accuracy of LHCb event simulation and the power of the $FD_{[z_D^{vtx} - z_B^{vtx}]}$ cut on reducing non-D events. Furthermore, the level of charmless background found in data is fully compatible (or even lower) with the one expected from purely MC calculations and naive rescaling of branching fractions.

4.5 $B \rightarrow [hh]_D h$ Internal Cross Feed

This section investigates possible cross feed between all the possible $B \rightarrow [hh]_D h$ modes, e.g. genuine $B \rightarrow [hh]_D h$ events reconstructed as signal events but under a different mass hypothesis (thus falling in the wrong $[hh]_D$ slice of the fitter, see Chap. 5). Cross feed between the different signal modes may occur because the particle identification of the kaons and the pions is imperfect. It is a peaking background that, unless properly controlled, can bias the *CP* observables. $B \rightarrow D\pi$

Table 4.9 Number of expected charmless events for each mode considered with respect to a given $FD_{[z_D^{vtx}-z_B^{vtx}]}$ cut in the offline selected sample

$FD_{[z_D^{vtx}-z_B^{vtx}]} >$	$B^+ \to K^+K^-K^+$	$B^+ \to K^+K^-\pi^+$	$B^+ \to \pi^+\pi^-K^+$	$B^+ \to \pi^+\pi^-\pi^+$
No cut	404.3 ± 27.2	$*60$	257.4 ± 19.8	238.7 ± 21.9
0.0	212.5 ± 21.5	$*32$	134.8 ± 15.1	113.8 ± 16.1
1.0	103.9 ± 17.1	$*15$	53.1 ± 12.9	48.7 ± 10.9
2.0	27.3 ± 12.3	**$*4$**	**12.5 ± 6.1**	**13.1 ± 8.6**
3.0	10.9 ± 10.5	$*2$	0.0 ± 1.8	3.9 ± 7.0
4.0	3.9 ± 9.4	$*1$	0.0 ± 1.3	0.0 ± 25.4

The first two columns are obtained from fits on the KK sideband, the second columns from the $\pi\pi$ sideband (uncertainties are statistical only). $B^+ \to K^+K^-\pi^+$ yields (marked with an asterisk) are derived from $B^+ \to K^+K^-K^+$ rescaling the measured corresponding branching fraction. Fits after $FD_{[z_D^{vtx}-z_B^{vtx}]} > 2$ are not statistically relevant and the signal-like peaking structure in the data completely vanishes

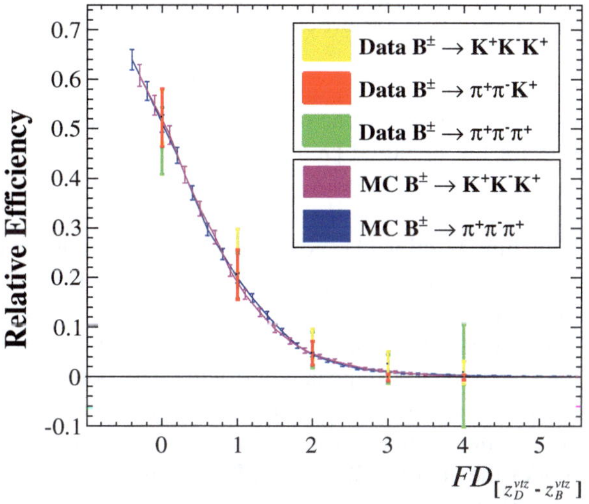

Fig. 4.28 Comparison of MC/data dependence of charmless yields with respect to the $FD_{[z_D^{vtx}-z_B^{vtx}]}$ cuts. Efficiencies are normalized to 1 when no $FD_{[z_D^{vtx}-z_B^{vtx}]}$ cut is applied. Excellent agreement is found between data and MC simulation

events where the bachelor track has been misidentified as a kaon (but the D meson is correctly reconstructed) are treated separately in Sect. 3.9 as no $\Delta LL_{K/\pi}$ cut is applied on the bachelor track and thus those events are considered part of the signal.

4.5.1 Misidentification of the $[hh]_D$ Daughter Tracks

Contamination between the $[hh]_D$ modes can occur when one or both daughter tracks are identified with the wrong mass hypothesis, leading to tails in the D mass distribution that could leak into the D signal mass window. This process can be studied with MC simulations (and then checked on data) by reconstructing every signal sample under different $[hh]_D$ hypotheses and thus simulating the effect of the

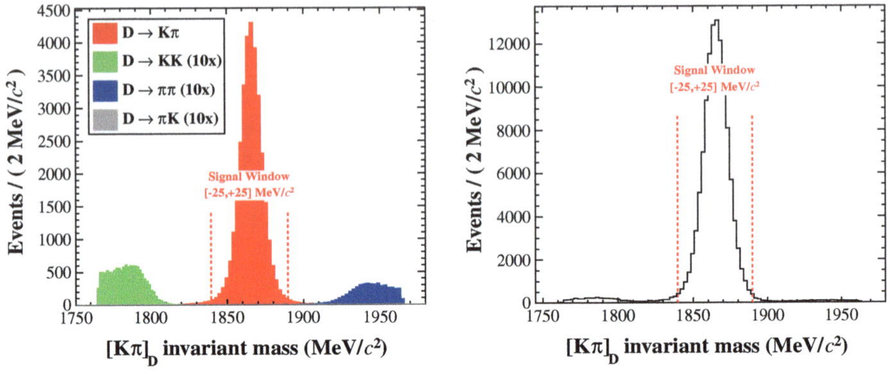

Fig. 4.29 $[K\pi]_D$ invariant mass (preliminary selection only and no PID requirement applied) for $B \to [hh]_D h$ events. *Left* MC truth matched events; every different $D \to hh$ contribution has been highlighted with a different colour, while $D \to KK, \pi\pi, \pi K$ contributions have been increased by a factor of 10 to improve clarity. *Right* data sample

misidentification of one or more daughter particles. The resulting D mass distribution for the favoured $[K\pi]_D$ mode is plotted in Fig. 4.29, for the $[KK, \pi\pi]_D$ *CP* modes in Figs. 4.30 and 4.31, while for the suppressed $[\pi K]_D$ mode in Fig. 4.32. MC Simulated events are naively normalized according to the measured $D \to hh$ sub-branching fraction [1] and the number of generated events for each sample. The same pattern of shifts is found in every plot: every time a pion is reconstructed as a kaon the distribution is shifted to the left part of the mass range with the expected left tail, while every time a kaon is misidentified as a pion the contrary applies. For this reason, the favoured $[K\pi]_D$ peak is surrounded by both sides by misidentified $B \to [hh]_D h$ events, while $[KK]_D$ and $[\pi\pi]_D$ peaks have misidentified signal events only in one side of their D mass distributions. As shown in the plots, in the favoured and the *CP* modes the cut of $[-25, +25]$ MeV/c^2 from the nominal D^0 mass [1] removes every possible contamination from internal cross feed between the different $[hh]_D$ slices. However, a similar argument cannot be applied to the $[\pi K]_D$ suppressed mode: in this case, the double-misidentification of both daughter tracks of the favoured decay mode plays a major role in its expected background level and thus this effect must be taken into account very accurately. The favoured → suppressed-mode cross feed is studied separately in Sect. 4.5.3.

4.5.2 Internal Bachelor-Daughter Combinatorial Background

Internal cross feed can derive from $B \to [hh]_D h$ signal modes which have the same hhh final state, where the three tracks are correctly identified and originate from the same signal B meson, but the intermediate $D \to hh$ candidate is not reconstructed correctly. This occurs when $B^+ \to [dau^+ dau^-] bach^+$ decays are reconstructed as $B^+ \to [bach^+ dau^-] dau^+$, e.g one of the daughter and the bachelor tracks

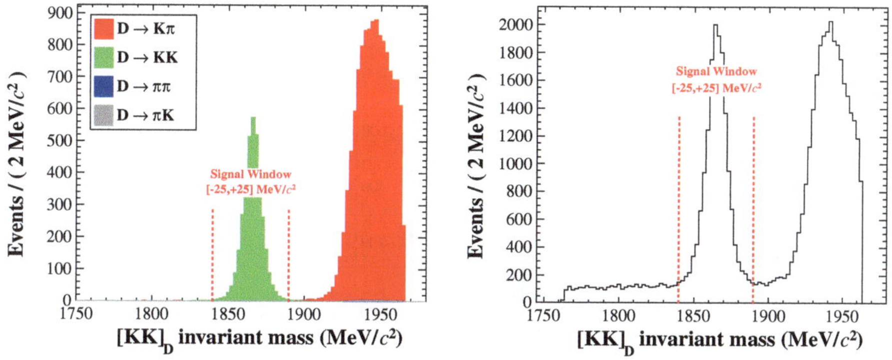

Fig. 4.30 $[KK]_D$ invariant mass (preliminary selection only and no PID requirement applied) for $B \to [hh]_D h$ events. *Left* MC truth matched events; every different $D \to hh$ contribution has been highlighted with a different colour. *Right* data sample

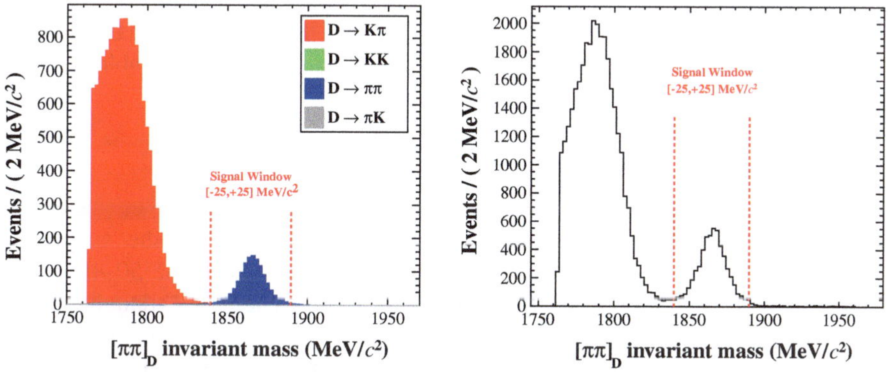

Fig. 4.31 $[\pi\pi]_D$ invariant mass (preliminary selection only and no PID requirement applied) for $B \to [hh]_D h$ events. *Left* MC truth matched events; every different $D \to hh$ contribution has been highlighted with a different colour. *Right* data sample

are swapped between each other. Since the final state is the same in both cases, these candidates have a peaking B mass distribution with same mean and width of genuine signal candidates, but they are expected to have no structures in the D mass. The cuts on m_D and $FD_{[z_D^{vtx}-z_B^{vtx}]}$ help reducing this peaking background but there might be some remaining contribution when the branching fraction of the cross feed mode is particularly high. This is true in the $B^+ \to [\pi^+\pi^-]_D K^+$ case, where the favoured and most abundant $B^+ \to [K^+\pi^-]_D \pi^+$ mode can be a background, with a branching fraction about 360 times higher. This case is shown in Fig. 4.33 for MC simulated events and data. Internal bachelor-daughter combinatorial background is suppressed introducing a mass veto on the neutral combination of the bachelor track and one of the daughter tracks. Only one $[dau^\mp, bach^\pm]$ combination is possible besides the signal $[dau^\pm, dau^\mp]_D$ combination and candidates are required to have

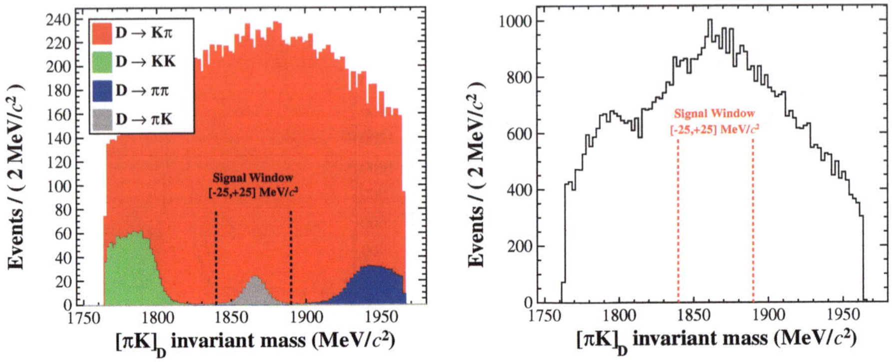

Fig. 4.32 $[\pi K]_D$ invariant mass (preliminary selection only and no PID requirement applied) for $B \to [hh]_D h$ events. *Left* MC truth matched events; every different $D \to hh$ contribution has been highlighted with a different colour. *Right* data sample

$m_{[dau^{\mp}bach^{\pm}]} \notin [-22, +22]\,\text{MeV/c}^2$ window around the nominal D mass [1]. This veto is applied to every $[hh]_D$ slice in order not to bias the CP observables.

4.5.3 Favoured → Suppressed-Mode Cross Feed

$B \to [hh]_D h$ internal cross feed is particularly important for the suppressed $B \to [\pi K]_D K$ mode, whose branching fraction is expected to be 50–100 times smaller than that of the favoured $B \to [K\pi]_D K$ mode. When the D daughter's PID assignment is swapped ($K \leftrightarrow \pi$), as would be the case when the favoured mode is reconstructed as the suppressed, the invariant mass distribution peaks at the B mass although with a width larger than that of the genuine signal (see Fig. 4.32). In this section, a strategy to reduce the cross feed with PID cuts on the D daughter tracks and a veto on the swapped invariant mass sum is described.

Mass Spectra with Wrong Mass Hypothesis

The invariant mass distribution of D candidates after a swap of their daughters' mass hypotheses is a wide distribution centred at the D^0 mass. The double misidentification of the two daughter tracks, in fact, does not shift the mean of the D distribution but it increases its width to a range of about 1200 MeV/c^2. This feature can be used to easily remove most of this background just applying a veto on the *double-swapped* D mass variable, e.g the $[K\pi]_D$ mass computed using the same daughter tracks but exchanging their mass assignments. The proportion of this distribution surviving any D mass selection cut can be calculated from data using the abundant statistics from the favoured mode once the full offline selection is applied. The distribution of swapped-mass-hypothesis D candidates ($K \leftrightarrow \pi$) is shown in Fig. 4.34 and can be

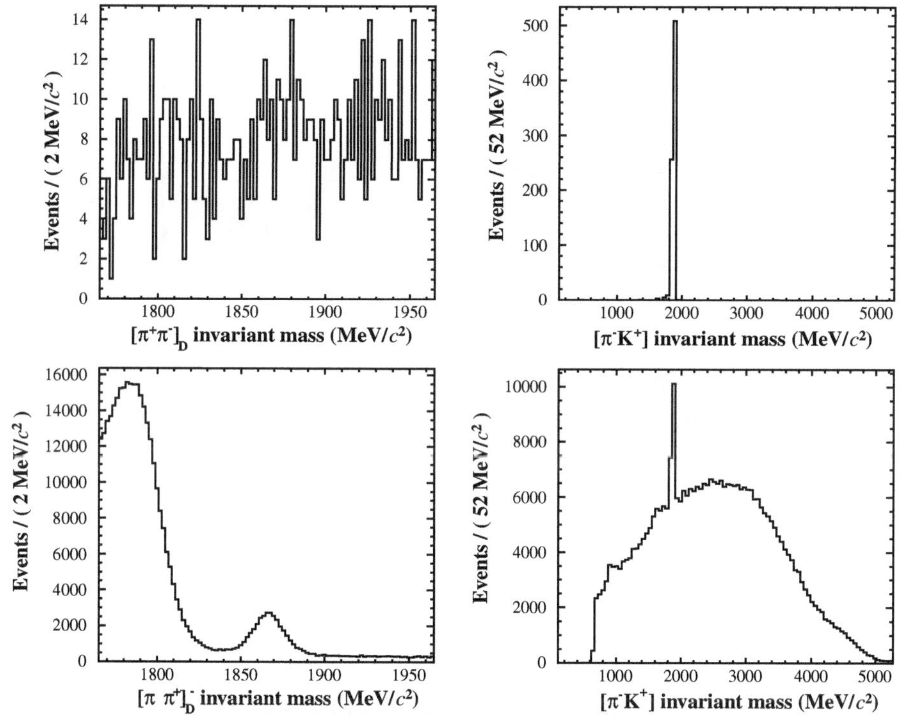

Fig. 4.33 Invariant masses of both two neutral $[h^+h^-]$ combinations for $B \to [\pi\pi]_D K$ candidates. *Top* favored MC simulated $B \to [K\pi]_D\pi$ events reconstructed under the $B \to [\pi\pi]_D K$ mass hypothesis (same final state but the D candidate is formed with the wrong $[h^+h^-]$ combination). *Bottom* $B \to [\pi\pi]_D K$ candidates reconstructed in data

Table 4.10 Retention of the favoured → suppressed cross feed veto after the full offline selection is applied

Veto $\|m_D - m_0\| >$	10 MeV/c^2 (%)	15 MeV/c^2 (%)	20 MeV/c^2 (%)	25 MeV/c^2 (%)
Cross feed retention	25.2	10.8	4.9	2.4
Signal retention	94.6	92.0	89.3	86.6
Cross feed retention	23.9	9.7	4.1	1.9
Signal retention	94.3	91.7	89.1	86.5

The top rows show the obtained efficiencies for $\Delta LL^{bach}_{K/\pi} < 4$ on the bachelor track, the bottom rows the same efficiencies for $\Delta LL^{bach}_{K/\pi} > 4$

easily compared with the narrow true D peak. In both plots, vertical dashed red lines indicate a possible veto window. The proportion of the events inside this window is removed.

From inspecting the integrals of the D peaks, both before and after the swap of the daughter track masses, the proportion of events surviving any D mass veto is measured with this totally data-driven approach and shown in Table 4.10, split by

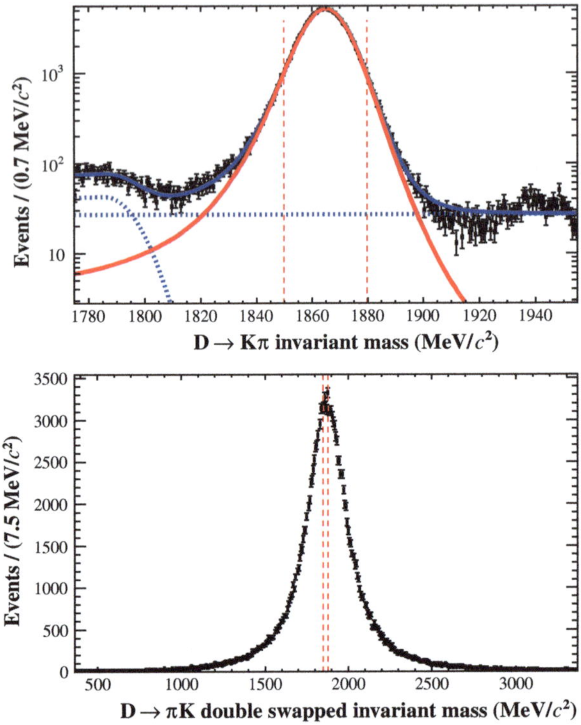

Fig. 4.34 *Top* fit to the D peak from the favored $[K\pi]_D$ mode in data. A modified gaussian is used to model the peak and its non-Gaussian tails (see Sect. 3.9). From this fit, the peak is isolated using sWeights [9] which are used to give a distribution of D candidates where the mass hypothesis of the daughters has been swapped. sWeights use the known correctly reconstructed D distribution (*top*) to isolate the signal *double-swapped* component from the other sources of backgrounds (*bottom*). *Bottom* the obtained *double-swapped* (daughter $K \leftrightarrow \pi$) mass distribution using the sWeights technique. The applied veto of ±15 MeV/c^2 is indicated with the vertical *dashed lines*

$\Delta LL^{bach}_{K/\pi}$ cut (4 is the chosen value in Chap. 5). A veto window of 15 MeV/c^2 (two σ of the peak width) is chosen. It will remove (90.3 ± 0.3) % of favoured candidates that cross feed into the suppressed mode at the cost of removing (8.3 ± 0.3) % suppressed $B \rightarrow [\pi K]_D K$ events.

Daughter $\Delta LL_{K/\pi}$ Cut Efficiencies

In order to remove further the favoured → suppressed-mode cross feed, cuts are applied on the PID $\Delta LL_{K/\pi}$ variables of the daughter tracks. The PID efficiencies for the D daughter tracks can be measured directly on the favoured $B \rightarrow [K\pi]_D h$ data sample by repeating the fit on the $[K\pi]_D$ invariant mass spectrum for a variety of PID $\Delta LL_{K/\pi}$ cuts. The result of this procedure is shown in Table 4.11 for the genuine signal candidates and in Table 4.12 for the cross feed events. The same

Table 4.11 Efficiency for $B \to [\pi K]_D K$ signal candidates for combined $\Delta LL_{K/\pi}$ PID cuts on the D daughter tracks

%	<200	<4	<2	<0	<-2	<-4	<-6	<-8
>-200	100	94.38 ± 0.23	91.56 ± 0.28	85.70 ± 0.34	81.09 ± 0.39	75.78 ± 0.43	71.03 ± 0.46	66.54 ± 0.47
>0	95.76 ± 0.20	90.27 ± 0.30	87.47 ± 0.33	82.77 ± 0.38	77.60 ± 0.42	72.56 ± 0.45	68.01 ± 0.47	63.73 ± 0.48
>2	91.55 ± 0.28	86.33 ± 0.35	83.67 ± 0.37	79.34 ± 0.41	74.56 ± 0.44	69.93 ± 0.46	65.60 ± 0.48	61.49 ± 0.49
>4	88.11 ± 0.33	83.08 ± 0.38	80.56 ± 0.40	76.46 ± 0.43	71.98 ± 0.45	67.48 ± 0.47	63.34 ± 0.48	59.45 ± 0.49
>6	84.48 ± 0.36	79.59 ± 0.41	77.19 ± 0.42	73.36 ± 0.44	69.08 ± 0.46	64.88 ± 0.48	60.88 ± 0.49	57.16 ± 0.50
>8	80.83 ± 0.40	76.11 ± 0.43	73.90 ± 0.44	70.24 ± 0.46	66.15 ± 0.48	62.16 ± 0.49	58.42 ± 0.50	54.84 ± 0.50

Rows indicate the chosen $\Delta LL_{K/\pi}$ cut on the daughter pion track (less than), while columns indicate $\Delta LL_{K/\pi}$ cuts on the daughter kaon track (greater than)

Table 4.12 Efficiency for favoured → suppressed-mode cross feed candidates for combined $\Delta LL_{K/\pi}$ PID cuts on the D daughter tracks

%	<200	<4	<2	<0	<−2	<−4	<−6	<−8
>−200	100	12.70 ± 0.34	9.26 ± 0.29	5.03 ± 0.22	2.32 ± 0.15	1.14 ± 0.11	0.77 ± 0.09	0.60 ± 0.08
>0	13.65 ± 0.35	2.25 ± 0.15	1.66 ± 0.13	0.80 ± 0.09	0.37 ± 0.06	0.15 ± 0.04	0.11 ± 0.03	0.09 ± 0.03
>2	8.60 ± 0.28	1.24 ± 0.11	0.88 ± 0.09	0.45 ± 0.07	0.28 ± 0.05	0.16 ± 0.04	0.14 ± 0.04	0.11 ± 0.03
>4	5.71 ± 0.23	0.82 ± 0.09	0.60 ± 0.08	0.32 ± 0.06	0.19 ± 0.04	0.12 ± 0.03	0.12 ± 0.03	0.10 ± 0.03
>6	3.70 ± 0.19	0.53 ± 0.07	0.38 ± 0.06	0.21 ± 0.05	0.11 ± 0.03	0.03 ± 0.02	0.04 ± 0.02	0.03 ± 0.02
>8	2.36 ± 0.15	0.34 ± 0.06	0.23 ± 0.05	0.14 ± 0.04	0.05 ± 0.02	0.00 ± 0.00	0.00 ± 0.00	0.00 ± 0.01

Rows indicate the chosen $\Delta LL_{K/\pi}$ cut on the daughter pion track (less than), while columns indicate $\Delta LL_{K/\pi}$ cuts on the daughter kaon track (greater than). In the cross feed events, the mass assignment of the daughter tracks has been swapped ($K \leftrightarrow \pi$)

cuts on $\Delta LL_{K/\pi}$ are applied to both favoured and suppressed modes (and the *CP* modes for consistency) in order not to bias the *CP* observables. Mutually-exclusive choices exclude the possibility of the same candidate appearing in both favoured and suppressed final selections. For example, from the tables, the proportion of favoured $B \to [K\pi]_D h$ candidates being misidentified is (0.28 ± 0.05) % for a cut of $\Delta LL^{Kdau}_{K/\pi} > 2$ and $DLK^{\pi dau}_{K} < -2$.

PDF Parametrization and Estimation of the Cross Feed Rate

The distribution of cross feed events is studied on MC simulated $B \to [K\pi]_D h$ events reconstructed as suppressed $B \to [\pi K]_D h$: the simulated events surviving the veto are found to peak at the nominal B mass, but they are spread on a wider distribution. In the signal extraction fit, those events are modeled with a Gaussian distribution, centred at the measured B mass value, but 2.1 times larger than the signal component, as shown in Fig. 4.35. Similarly, cross feed events with the bachelor misidentified are assigned a PDF identical to that of the $B \to D\pi$ misidentified signal, but with its width parameter inflated by a factor two. The favoured $\to$ suppressed cross feed rate is calculated in Table 4.13, assuming no correlation between the cuts on $(m_D, m_{D(K\leftrightarrow\pi)})$ and $\Delta LL^{dau}_{K/\pi}$. Under this assumption, the relative rate to the favoured mode is $(6 \pm 2) \times 10^{-5}$ for a veto of $>15\,\text{MeV}/c^2$ on the *double-swapped*

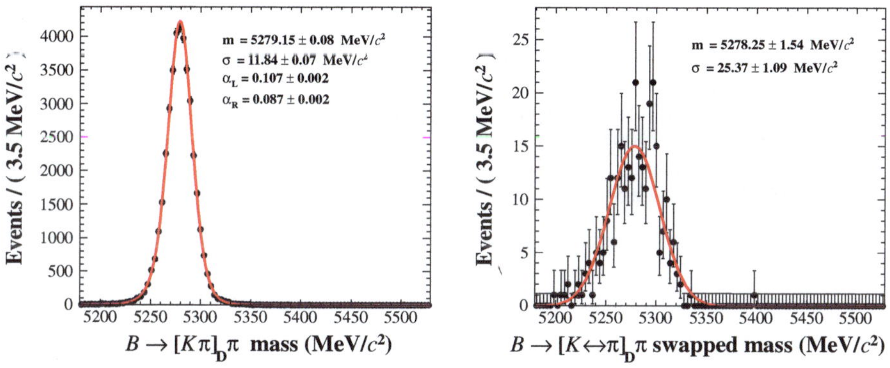

Fig. 4.35 *Left* genuine signal $B \to [K\pi]_D\pi$ events reconstructed with the correct D daughter mass hypotheses. *Right* $B \to [K\pi]_D\pi$ events reconstructed under the $B \to [\pi K]_D\pi$ suppressed hypothesis in order to simulate the $K \leftrightarrow \pi$ double swap. All selections and vetoes are applied on the samples: the width of the *double-swapped* peak is found to be about 2 times larger then the signal one

Table 4.13 Response of the correct and incorrect mass hypotheses to the cuts designed to reduce the cross feed

	Cross feed (%)	Signal (%)
$\|m_D - \mu_D\| < 25$	(13.4 ± 0.3)	(97.6 ± 0.3)
$m_{D(K \leftrightarrow \pi)}$ veto	(9.7 ± 0.3)	(91.7 ± 0.3)
$\Delta LL^{\pi}_{K/\pi} < -2$, $\Delta LL^{K}_{K/\pi} > 2$	(0.3 ± 0.1)	(74.6 ± 0.4)
Total	(0.004 ± 0.001)	(66.7 ± 0.5)
Relative ratio	$(6 \pm 2) \times 10^{-5}$	

References

1. Particle Data Group, K. Nakamura et al., Review of particle physics. J. Phys. **G37** 07502 (2010)
2. D.J. Lange, The EvtGen particle decay simulation package. Nucl. Instrum. Meth. **A462** 152 (2001)
3. CDF Collaboration, T. Aaltonen et al., Measurement of ratios of fragmentation fractions for bottom hadrons in $p\overline{p}$ collisions at $\sqrt{s} = 1.96$ TeV. Phys. Rev. D **77** 072003 (2008)
4. The LHCb Collaboration, R. Aaji et al., Measurement of b hadron production fractions in 7 TeV pp collisions. Phys. Rev. D **85** 032008 (2012)
5. K. Cranmer, Kernel estimation in high-energy physics. Comput. Phys. Commun. **136**, 198 (2001). arXiv:hep-ex/0011057
6. LHCb Collaboration, R. Aaij et al, Measurements of the branching fractions and CP asymmetries of $B^+ \to J/\psi\pi^+$ and $B^+ \to \psi(2S)\pi^+$ decays. Phys. Rev. **D85** (2012) 091105. arXiv:1203.3592
7. LHCb Collaboration, R. Aaij et al., First observation of the decay $\bar{B}^0_s \to D^0 K^{*0}$ and a measurement of the ratio of branching fractions $\frac{(\bar{B}^0_s \to D^0 K^{*0})}{(\bar{B}^0 \to D^0 \rho^0)}$. Phys. Lett. **B706**, 32 (2011). arXiv:1110.3676
8. BABAR Collaboration, B. Aubert et al., Observation of the decay $B^+ \to K^+ K^- \pi^+$, Phys. Rev. Lett. **99**, 221801 (2007)
9. M. Pivk, F.R. Le Diberder, A statistical tool to unfold data distributions. Nucl. Instrum. Phys. Res. A **555**(12), 356 (2005)

Chapter 5
The Simultaneous Fit

This chapter presents a detailed description of the maximum likelihood (ML) fit used in the extraction of the *CP* physics observables. The general idea is to perform a simultaneous fit to the B invariant mass distributions of all the different $B \to [hh]_D h$ modes at the same time and take advantage of a large set of parameters in common. Most parameters can be left floating in the minimization process exploiting the high statistics of the most abundant $B \to [K\pi]_D\pi$ favoured control sample and thus allowing a better adjustment of the PDFs on data itself. The chapter is divided in three sections. Section 5.1 describes the fitter package implemented in the ROOFIT [1] framework, the general strategy for the extraction of the *CP* observables and the components used in each fit subsample. The final configuration of the fit has been validated with *toy MC studies*[1] which are presented in Sect. 5.2. The chapter concludes with a description of all the possible sources of systematic uncertainties and their impact on the physical observables. Every assumption adopted in the analysis, e.g differences in the kaon/pion selection, the PID efficiency evaluation and any fixed parameter in the PDF shapes, reflects into a systematic uncertainty on all the measured values. These systematic uncertainties are propagated through the final fit as described in Sect. 5.3. The numerical results from the fitter are presented in Chap. 6.

5.1 Fitter Strategy

5.1.1 Introduction

The observables of interest are extracted with an extended binned maximum likelihood fit to the invariant mass distributions of the selected B candidates. Sensitivity to *CP* asymmetries is achieved by separating the candidates by charge into B^- and

[1] Toy experiments are MC simulated experiments where every dataset is generated from the expected PDFs and then fitted with the same model used for data.

P. Gandini, *Observation of CP Violation in $B^\pm \to DK^\pm$ Decays*, Springer Theses, DOI: 10.1007/978-3-319-01029-8_5,

B^+ separate samples. For all the four $[K\pi, KK, \pi\pi, \pi K]_D$ modes, $B \to [hh]_D K$ and $B \to [hh]_D \pi$ decays can be separated to around 90 % purity by a requirement on the $\Delta LL^{bach}_{K/\pi}$ of the bachelor track. Events *passing* this cut are kaon-like and are reconstructed as $B \to DK$ (PASS subsample). Events *failing* the cut are reconstructed as the $B \to D\pi$ final state (FAIL subsample). Therefore, most $B \to DK$ candidates are found in the PASS subsample, while $B \to D\pi$ events dominate the FAIL subsample. For the final configuration of the fitter, the value $\Delta LL^{bach}_{K/\pi} = 4$ is chosen and the efficiency of this cut is determined from the data-driven calibration procedure described in Sect. 3.11. This value has been chosen since it provides a low pion misidentification rate, ϵ_π, of about 3.8 % (Tables 3.12 and 3.13) while ensuring a good signal efficiency for $B \to DK$ events of about 87 % (Tables 3.14 and 3.15). In fact, having a large $B \to D\pi$ misidentified component makes it more difficult to resolve correctly reconstructed $B \to DK$ signal events, while a too hard cut on $\Delta LL^{bach}_{K/\pi}$ forces too much $B \to DK$ events into the FAIL subsample, where they are drowned in the background and their information is lost. Note that the $\Delta LL^{bach}_{K/\pi}$ cut only splits events into the PASS and FAIL subsamples and no candidate is rejected by this requirement. The fit therefore comprises four subsamples $(B^-, B^+) \times$ (PASS, FAIL) for each $[hh]_D$ mode, fitted simultaneously with a set of parameters in common. This is sketched in Fig. 5.1 for a simulated toy sample of the $[K\pi]_D$ favoured subsample.

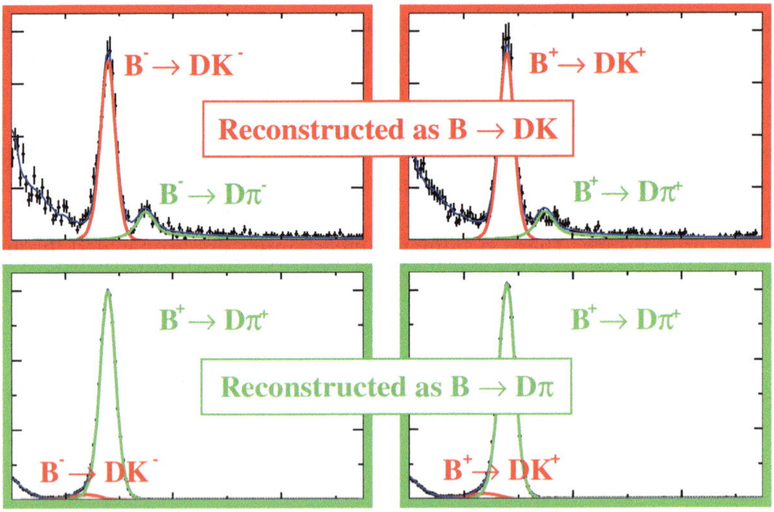

Fig. 5.1 Fitter configuration for a generated sample emulating the $B \to [K\pi]_D h$ data. *Left* $B^- \to Dh^-$ candidates. *Right* $B^+ \to Dh^+$ candidates. The *top row* contains candidates which pass the $\Delta LL^{bach}_{K/\pi} > 4$ PID requirement on the bachelor track (PASS subsample), reconstructed as $B \to DK$. The *bottom row* contains candidates with $\Delta LL^{bach}_{K/\pi} < 4$ (FAIL subsample), reconstructed as $B \to D\pi$. In this picture, the proportion of misidentified $B \to DK$ events in the FAIL subsample has been artificially increased by a factor of 10 to improve visibility, since these events would normally be drowned by the huge $B \to D\pi$ peak

The ML fit is performed simultaneously over the four D modes, so four sets of $(B^-, B^+)\times$(PASS, FAIL) plots are produced, one for each $D \to hh$ final state. The binned version of the fit is preferred since it is about five times faster compared with its unbinned version and degrades the errors of the fit parameters only in their third significant place. Bins of 5MeV$/c^2$ are chosen to cover the invariant mass range of $5100 - 5750\,\mathrm{MeV}/c^2$ (130 bins in total).

5.1.2 Fitter Components

The total PDF model is built from four or five components in each D subsample representing the various sources of events in each mode.

Correctly Reconstructed Signal $B \to [hh]_D h$ Components

The $B \to [hh]_D h$ signal lineshapes used in the fitter are described in Sect. 3.9. Correctly reconstructed $B \to [hh]_D K$ (in the PASS subsample) and $B \to [hh]_D \pi$ (in the FAIL subsample) are modeled with the modified Gaussian PDF defined in Sect. 3.9.1. The mean μ of each signal peaks is floated and required to be common in all the subsamples. The tail parameters α_L and α_R of the distribution are left floating as well, but they are common parameters for both $B^\pm \to D\pi^\pm$ and $B^\pm \to DK^\pm$ peaks. The width of the $B \to D\pi$ PDF floats independently for each $[hh]_D$ mode, though common between the favoured $[K\pi]_D$ and suppressed $[\pi K]_D$ modes, while each $\sigma_{B\to DK}$ width is related to that of the corresponding $B \to D\pi$ peak $\sigma_{B\to D\pi}$ by a multiplicative factor R_σ of 0.945, 0.950 and 0.955 for $[\pi\pi]_D$, $[K\pi]_D$ and $[KK]_D$ respectively. These assumptions are justified from simulation in Fig. 5.2.The central value of 0.950 is also found from a fit to the most abundant favoured peak once a hard PID cut is applied to reduce the misidentification contribution, as shown in Fig. 5.3.This fit finds the $\sigma_{B\to DK}/\sigma_{B\to D\pi}$ ratio to be 0.95 ± 0.02. This ratio is introduced in the final fitter configuration as a fixed number and its uncertainty is propagated when considering the systematics. The differences from this central value found in simulated events for the $[\pi\pi]_D$ mode (−0.5 %) and the $[KK]_D$ mode (+0.5 %) are taken into account in the final fitter and also carry an additional 0.5 % systematic.

Misidentified Signal $B \to [hh]_D h$ Components

The shape of the $B \to DK$ candidates, reconstructed as $B \to D\pi$ in the FAIL subsample is the sum of two Crystal Ball functions with all parameters fixed from simulation, as defined in Sect. 3.9.4. The shape of $B \to D\pi$ candidates reconstructed at $B \to DK$ is also a sum of two Crystal Ball functions with a common mean for all the D modes. All other shape parameters float in the fit except for the two parameters a and n that define the lower tail of the distribution. These are both fixed at (2 ± 1) and

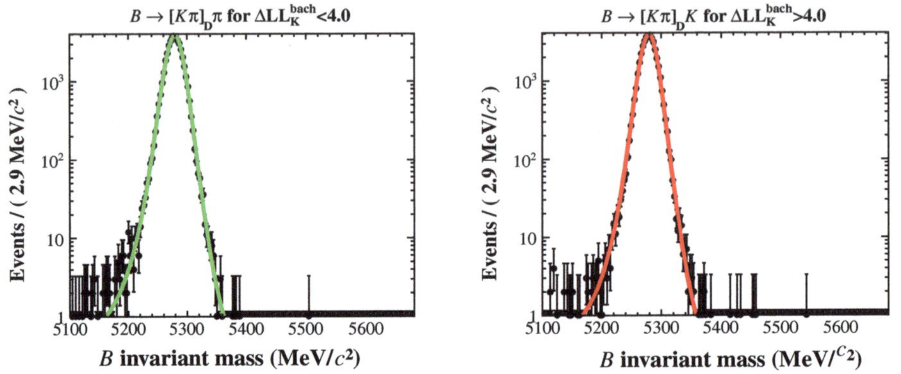

Fig. 5.2 A simultaneous fit to correctly reconstructed $B \to D\pi$ and $B \to DK$ simulated events after the full selection applied. $B \to D\pi$ candidates (*Left*) are required to satisfy $\Delta LL^{bach}_{K/\pi} \leq 4$, while $B \to DK$ candidates (*Right*) $\Delta LL^{bach}_{K/\pi} \geq 4$ in order to reproduce the PASS/FAIL fitter configuration described in Sec. 5.1.1. The mean and tail parameters are required to be the same. The $\sigma_{B\to D\pi}$ width is floated and $\sigma_{B\to DK} = R_\sigma \sigma_{B\to D\pi}$, where R_σ is extracted directly from the fit. r is found to be 0.945, 0.950 and 0.955 for $B \to [\pi\pi]_D h$, $B \to [K\pi]_D h$ and $B \to [KK]_D h$ simulated events respectively with a statistical uncertainty of ± 0.005

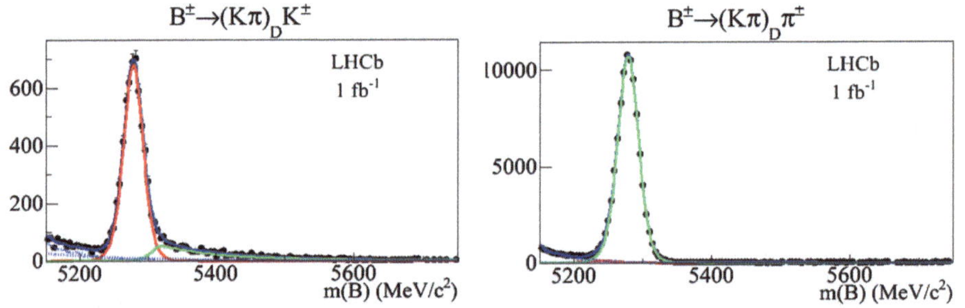

Fig. 5.3 A simultaneous fit to the favoured $[K\pi]_D$ sample with a $\Delta LL^{bach}_{K/\pi}$ cut at 8. The $B \to D\pi$ width is $(14.7 \pm 0.8)\text{MeV}/c^2$ and the $B \to DK$ peak is narrower by a factor of (0.953 ± 0.020). A similar fit with a $\Delta LL^{bach}_{K/\pi}$ cut at six gives (0.946 ± 0.018) for the ratio

the resulting fitted shape is used in (and defined simultaneously from) all D modes considered.

Combinatorial Background

Combinatorial background in the final sample is described in detail in Sect. 4.1. A linear approximation is adequate to describe the slope across the invariant mass spectrum considered, so a linear PDF is used for this background. A common slope parameter is used in all subsamples, though n_{comb} yields vary independently in each D mode.

Partially Reconstructed B $\rightarrow DX$

Backgrounds from partially reconstructed B $\rightarrow DX$ backgrounds are described in Sect. 4.3. The PDFs illustrated in Fig. 4.13 are used for all the D modes, with the important additions of:

- **Two extra $\Lambda_b \rightarrow [Kp\pi]_{\Lambda_c}h$ components for the $[KK]_D$ mode.** Though the total contribution is allowed to vary, the shape and relative proportion of $\Lambda_c^+ K^-$ and $\Lambda_c^+\pi^-$ components are fixed. Their relative proportion is unknown but assumed to be 0.060 ± 0.015, similar to the effective Cabibbo suppression observed in B mesons. The fitted $\Lambda_b \rightarrow [Kp\pi]_{\Lambda_c}\pi$ yield is consistent with the simulated expectations reported in Table 4.1. PDF in Fig. 4.16.
- **The $B_s \rightarrow [K\pi]_{D^0}[K\pi]$ component for the suppressed $[\pi K]_D$ mode.** In the fit, this yield is allowed to vary though the reported yield is in agreement with the simulated expectation, as derived in Table 4.2 PDF in Fig. 4.18.

The B^- and B^+ yields in each subsample vary independently, making no assumption of *CP* symmetry in the background.

5.1.3 Yields

B^+ and B^- yields for each fitter component are floated independently, both for signal and background. The fitter is thus stable against any possible *CP* violation in the background, though unexpected and unobserved. The number of signal $B^\pm \rightarrow Dh^\pm$ events is defined as:

$$n_{\text{TOT}}^{B^\pm \rightarrow D\pi^\pm} = n_{\text{PASS}}^{B^\pm \rightarrow D\pi^\pm} + n_{\text{FAIL}}^{B^\pm \rightarrow D\pi^\pm}$$
$$n_{\text{TOT}}^{B^\pm \rightarrow DK^\pm} = n_{\text{PASS}}^{B^\pm \rightarrow DK^\pm} + n_{\text{FAIL}}^{B^\pm \rightarrow DK^\pm}$$

n_{B^+} and n_{B^-} signal yields are then combined to construct the physics *CP* observables. The total number of $B \rightarrow [hh]_D\pi$ events $(n_{B^+} + n_{B^-})$ is a free parameter for each $[hh]_D$ subsample in the fit, while branching fraction ratios are extracted directly via:

$$R_{K/\pi}^{hh} = \frac{n_{\text{TOT}}^{B^\pm \rightarrow [hh]_D K^\pm}}{n_{\text{TOT}}^{B^\pm \rightarrow [hh]_D \pi^\pm}}$$

5.1.4 Charge Asymmetries

The measured charge asymmetries in each of the $B \rightarrow [hh]_D h$ signal modes considered are defined experimentally as:

$$A_{\text{raw}} = \frac{n_{B^-} - n_{B^+}}{n_{B^-} + n_{B^+}}$$

where $n_{B^-}$and $n_{B^+}$are the extracted number of signal events, split by charge. All asymmetries are normalized to $[-1, +1]$ by construction, with positive asymmetries indicating an excess of negatively charged B mesons. In addition to the weak *CP*-violation physical asymmetry A_{CP} , other charge asymmetries may be introduced by experimental factors:

- **Production asymmetry A_{prod}**
  This is the $B^{\mp}$ production asymmetry in pp collisions at $\sqrt{s} = 7\,\text{TeV}$. Since it is easier to produce B^+ $(\bar{b}u)$ than B^- $(b\bar{u})$ mesons in pp collisions,[2] a small negative production asymmetry is expected. $A_{\text{prod}}$is assumed to be $(-0.8 \pm 0.7)\,\%$ as deduced from the observed raw asymmetry of $B^{\pm} \to J/\psi K^{\pm}$ decays at LHCb [2].
- **Kaon detection asymmetry A_K^{det}** It is due to the different interaction lengths of K^+ and K^- mesons in the detector material. K^- mesons $(\bar{u}s)$ are expected to interact more with the detector material compared to K^+ mesons $(u\bar{s})$ so a small negative A_K^{det} is expected. A fixed value of $(-0.5 \pm 0.7)\,\%$ is assigned for every occurrence of strangeness in the $B \to [hh]_D h$ final state. As the total momentum dependence of A_K^{det} is assumed to be small in the full momentum range considered $(5 - 100\,\text{GeV}/c)$, no distinction is made between the asymmetry on the bachelor and the daughter tracks and the same A_K^{det} value is applied to both.
- **Pion detection asymmetry A_π^{det}**
  The equivalent detector asymmetry for pions is expected to be much smaller, compatible with zero. $A_\pi^{\text{det}} = (0.0 \pm 0.7)\%$ is assigned. This number has been confirmed in [3], where the LHCb detector A_π^{det} is measured to be $(0.08 \pm 0.25)\,\%$. The pion detection efficiencies are also found to be consistent with being independent of p and p_T. Note the conservative $\pm 0.7\,\%$ error assigned on A_π^{det} (compared to the $0.25\,\%$ error as in [3]).
- **Left/Right asymmetry $A_{L,R}^{\text{det}}$**
  $A_{L,R}^{\text{det}}$ is defined as the difference in the data-collection efficiency on the left and right part of the detector. In this case, any asymmetry in when comparing the left and right sides of the detector cancels out due to the approximate equal size of the two data samples. When fitting asymmetries separately for Magnet Up and Magnet Down samples, differences $< 0.5\,\%$ are observed compared to those quoted in this analysis. This systematic effect is thus included in the conservative uncertainty associated to A_π^{det} and it is not added separately.

The production asymmetry $A_{\text{prod}}$is common to all $B \to [hh]_D h$ modes but the kaon/pion total detection asymmetry content is different for the *CP* and $K\pi$ modes and different for $B \to DK$ and $B \to D\pi$ decays. This is because it depends upon the total number of kaons and pions in the final state. The physics *CP* asymmetries are defined as:

---

[2] $\bar{u}$ quarks must be extracted from the quark sea in the proton.

$$
\begin{aligned}
A_{\pi}^{K\pi} &= A_{CP}^{B\to[K\pi]_D\pi} &= A_{\rm raw}^{B\to[K\pi]_D\pi} - A_{\rm prod} - A_K^{\rm det}\\
A_{K}^{K\pi} &= A_{CP}^{B\to[K\pi]_D K} &= A_{\rm raw}^{B\to[K\pi]_D K} - A_{\rm prod} - 2A_K^{\rm det}\\
A_{\pi}^{KK} &= A_{CP}^{B\to[KK]_D\pi} &= A_{\rm raw}^{B\to[KK]_D\pi} - A_{\rm prod}\\
A_{K}^{KK} &= A_{CP}^{B\to[KK]_D K} &= A_{\rm raw}^{B\to[KK]_D K} - A_{\rm prod} - A_K^{\rm det}\\
A_{\pi}^{\pi\pi} &= A_{CP}^{B\to[\pi\pi]_D\pi} &= A_{\rm raw}^{B\to[\pi\pi]_D\pi} - A_{\rm prod}\\
A_{K}^{\pi\pi} &= A_{CP}^{B\to[\pi\pi]_D K} &= A_{\rm raw}^{B\to[\pi\pi]_D K} - A_{\rm prod} - A_K^{\rm det}
\end{aligned}
$$

where $A_{\rm raw}$ is the raw charge asymmetry measured on data and $A_{\pi}^{\rm det} = 0$ for clarity (but a similar argument applies). All six physics asymmetries are left floating, while $A_{\rm prod}$, $A_K^{\rm det}$ and $A_{\pi}^{\rm det}$ are fixed to the values indicated above with its systematic uncertainties associated to every assumption. The asymmetries for the suppressed $B \to [\pi K]_D h$ modes:

$$
\begin{aligned}
A_{\pi}^{\pi K} &= A_{CP}^{B\to[\pi K]_D\pi} = A_{\rm raw}^{B\to[\pi K]_D\pi} - A_{\rm prod} + A_K^{\rm det}\\
A_{K}^{\pi K} &= A_{CP}^{B\to[\pi K]_D K} = A_{\rm raw}^{B\to[\pi K]_D K} - A_{\rm prod}
\end{aligned}
$$

are not fitted directly, but the $R_K^-$and R_K^+ convention is preferred since it provides statistically independent variables (see Sect. 1.6.3). The charge asymmetries for the suppressed modes can be easily derived from the fitted $R_K^{\pm}$ and $R_{\pi}^{\pm}$ ratios in the following way:

$$
\begin{aligned}
A_{\rm ADS(\pi)} &= A_{\pi}^{\pi K} = \frac{R_{\pi}^- - R_{\pi}^+}{R_{\pi}^- - R_{\pi}^+}\\
A_{\rm ADS(K)} &= A_{K}^{\pi K} = \frac{R_{K}^- - R_{K}^+}{R_{K}^- - R_{K}^+}
\end{aligned}
$$

where the ADS notations have been introduced for consistency with Sect. 1.6.3

5.2 Validation

The performance and the stability of the fitter is tested using toy experiments. Toy experiments are MC simulated experiments where every dataset is generated from a given set of initial parameters and then fitted with the same model used for data. Initial parameters are chosen to resemble data as close as possible. With toy experiments it is possible to check the robustness of the fitter, its expected convergence rate and its ability to extract the physics observables, along with their expected experimental errors. The behaviour of any floating parameter introduced in the fitter can be double-checked as well. Pull distributions of each floating parameter θ introduced in the fit can be studied to evaluate any bias on the fitted quantities, their correct error estimation and any possible non-Gaussian behaviour in the distribution of the results. Pulls are defined as:

$$P(\theta) = \frac{\theta_{\text{fitted}} - \theta_{\text{generated}}}{\sigma_{\text{fitted}}}$$

They are expected to be distributed according to a Gaussian distribution with $\mu = 0$ and $\sigma = 1$. Means not compatible with zero indicate biases in the parameter extraction. Widths different from unity suggest undercoverage or overcoverage of the quoted error. Excluding non converged fits (2 % of the cases), the pull distributions of the most interesting quantities for the simultaneous fitter configuration presented in Sect. 5.1 are shown in Fig. 5.4 (about 1000 toy experiments generated). All pull distributions demonstrate a good stable behaviour of the fit and no large biases are found.

Furthermore, using these toy experiments, the expected fit correlations between the thirteen physics observables are shown in Fig. 5.5. No significant correlations are observed. This is not unexpected as the number of assumptions that could impact and correlate the physics measurements are kept to minimum. The origins of possible correlations would be:

- The same partially reconstructed background shape is used in all the subsamples,
- The same combinatoric linear slope parameter is used in all the subsamples, both for $B \to DK$ and $B \to D\pi$ modes,
- The ratio of the signal peak widths, $\sigma_{B \to DK}/\sigma_{B \to D\pi}$, is fixed at 0.95 for all the modes.

All of these have systematics assigned in Sect. 5.3.

5.3 Systematic Uncertainties

In this section, the various systematic biases and uncertainties that may affect the observable measurements are discussed. It should be noted that since all the measurements are ratios between $B \to [hh]_D h$ modes with exactly the same topology, many of the systematics cancel and only residual effects have to be taken into account. In particular, for the $R_{\text{ADS}(\pi)}$and the $R_{\text{ADS}(K)}$extraction, ratios are calucated between modes with exactly the same final state (except the relative charge of the particles) and most uncertainties cancel completely. The fit is designed to use the symmetry of the data sample to the maximum, making a minimum number of assumptions in the process. All the sources of systematic uncertainties are examined individually and their effect is propagated to the physical observables.

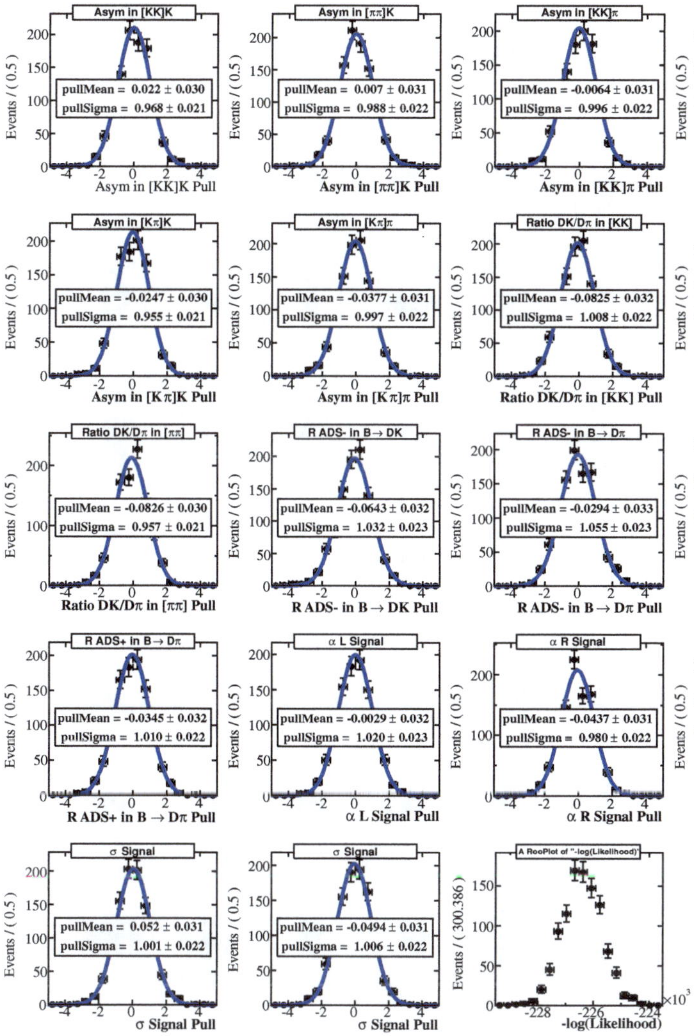

Fig. 5.4 Pull distributions for the most interesting variables from the full fit (the 13 *CP* observables and the signal peak parameters). The ΔLL distribution is also plotted

5.3.1 Multiplicative Corrections on $R^{hh}_{K/\pi}$

Although $B \to [hh]_D K$ and $B \to [hh]_D \pi$ decays have the same topology and the same $[hh]_D$ final state, small corrections (order of percent relative) are introduced to take into account possible small differences in the geometrical detector acceptance $\varepsilon^{\rm Geom}_{\rm eff}$, the selection efficiency (BDT cut plus background vetoes) $\varepsilon^{\rm Sel}_{\rm eff}$ and the trigger requirement efficiency $\varepsilon^{\rm Trig}_{\rm eff}$, due to the different K/π bachelor track. MC generated

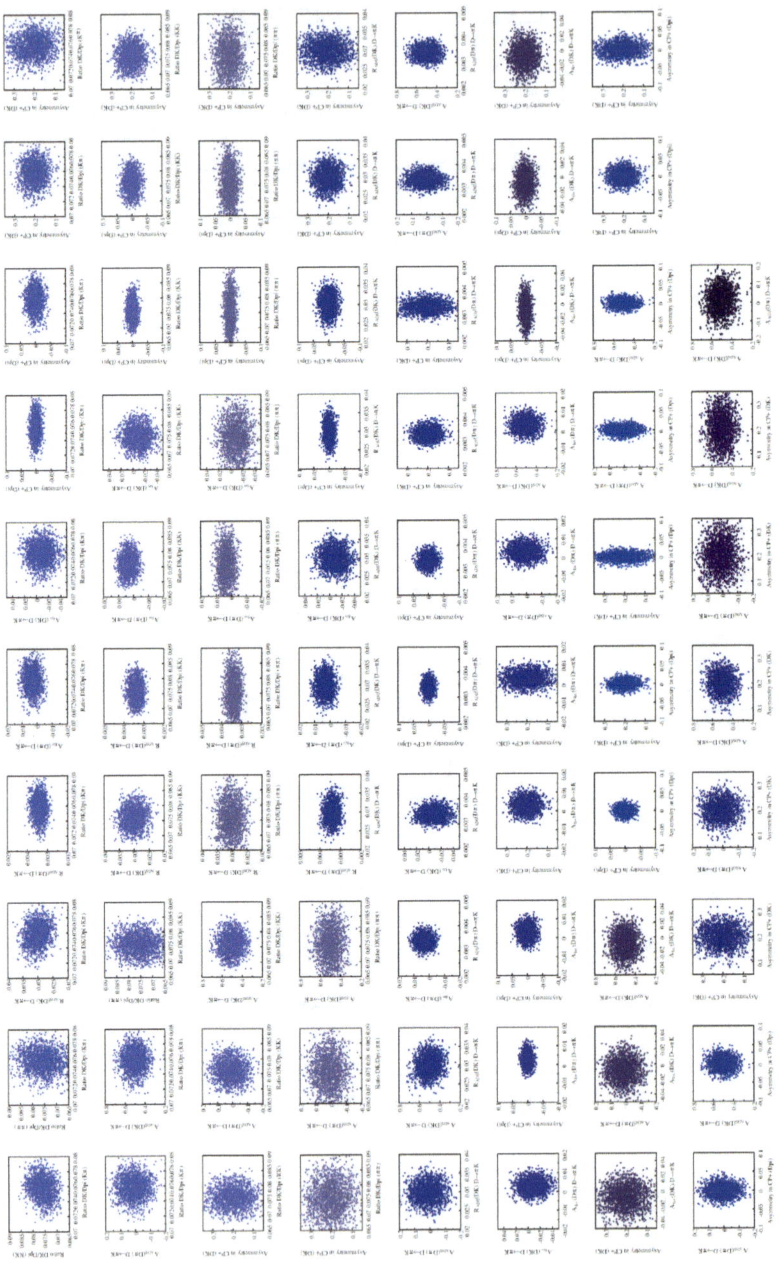

Fig. 5.5 Correlations amongst the physics variables from toy experiments. No obvious correlation is found

signal events are used to compute the relative correction factors and the ratio of branching fractions is corrected via:

$$R^{hh}_{K/\pi} = \frac{n_{B\to[hh]_D K}}{n_{B\to[hh]_D\pi}} = \frac{\mathcal{B}_{(B\to[hh]_D K)}}{\mathcal{B}_{(B\to[hh]_D\pi)}} \cdot C^{\text{Geom}}_{K/\pi} \cdot C^{\text{Sel}}_{K/\pi} \cdot C^{\text{Trig}}_{K/\pi}$$

where the correction factors C are defined as

$$C^{\text{Geom}}_{K/\pi} = \frac{\varepsilon^{\text{Geom}}_{\text{eff}(B\to[hh]_D K)}}{\varepsilon^{\text{Geom}}_{\text{eff}(B\to[hh]_D\pi)}},\quad C^{\text{Sel}}_{K/\pi} = \frac{\varepsilon^{\text{Sel}}_{\text{eff}(B\to[hh]_D K)}}{\varepsilon^{\text{Sel}}_{\text{eff}(B\to[hh]_D\pi)}} \quad\text{and}\quad C^{\text{Trig}}_{K/\pi} = \frac{\varepsilon^{\text{Trig}}_{\text{eff}(B\to[hh]_D K)}}{\varepsilon^{\text{Trig}}_{\text{eff}(B\to[hh]_D\pi)}}$$

Using this formula, a multiplicative factor for each contribution on the ratio is applied and a systematic uncertainty is assigned from the uncertainty of the correction. The LHCb detector geometrical acceptances are taken from Table 3.3 for the different $B \to [hh]_D h$ signal modes and $C^{\text{Geom}}_{K/\pi}$ factors are summarized in Table 5.1.The selection $\varepsilon^{\text{Sel}}_{\text{eff}}$ and TIS-TOS $\varepsilon^{\text{Trig}}_{\text{eff}}$ efficiencies are taken from Sect. 3.8 for all the modes. Efficiencies are taken from simulated events. In particular, any difference in the BDT selection between modes with a bachelor pion or a bachelor kaon has been investigated, since the bachelor p_{T} is used as an input variable for the BDT training (Sect. 3.5.1). The dependence of the total correction factor with respect to the BDT cut is shown in Fig. 5.6. A good stability is found over the whole region of interest.

5.3.2 Systematic Error on the Bachelor PID Variable

The efficiencies of kaon identification $\epsilon^{\text{PID}}_{K}$ and the pion misidentification rates $\epsilon^{\text{PID}}_{\pi}$are determined using the data-driven method described in Sect. 3.11. The calibration procedure uses a large $D^{*+} \to D^0(K^-\pi^+)\pi^+$ calibration sample and assumes that PID variables depend on the momentum p of the track, its η and the total number of tracks in the event. The obtained central values are $\epsilon^{\text{PID}}_{K} = 87.2\,\%$ and $\epsilon^{\text{PID}}_{\pi} = 3.8\,\%$ for the chosen $\Delta LL^{bach}_{K/\pi} > 4$ cut. The errors obtained with this

Table 5.1 Geometrical acceptance correction factors $C^{\text{Geom}}_{K/\pi}$

	$\varepsilon^{\text{Geom}}_{\text{eff}}$	$\varepsilon^{\text{Geom}}_{\text{eff}(B\to[hh]_D K)}/\varepsilon^{\text{Geom}}_{\text{eff}(B\to[hh]_D\pi)}$	Syst.
$B \to [K\pi]_D\pi$	0.1604 ± 0.0002	$C^{\text{Geom}}_{K/\pi} = 1.022 \pm 0.002$	(0.17 % rel.)
$B \to [K\pi]_D K$	0.1639 ± 0.0002		
$B \to [KK]_D\pi$	0.1676 ± 0.0002	$C^{\text{Geom}}_{K/\pi} = 1.023 \pm 0.002$	(0.17 % rel.)
$B \to [KK]_D K$	0.1714 ± 0.0002		
$B \to [\pi\pi]_D\pi$	0.1540 ± 0.0002	$C^{\text{Geom}}_{K/\pi} = 1.021 \pm 0.002$	(0.18 % rel.)
$B \to [\pi\pi]_D K$	0.1572 ± 0.0002		

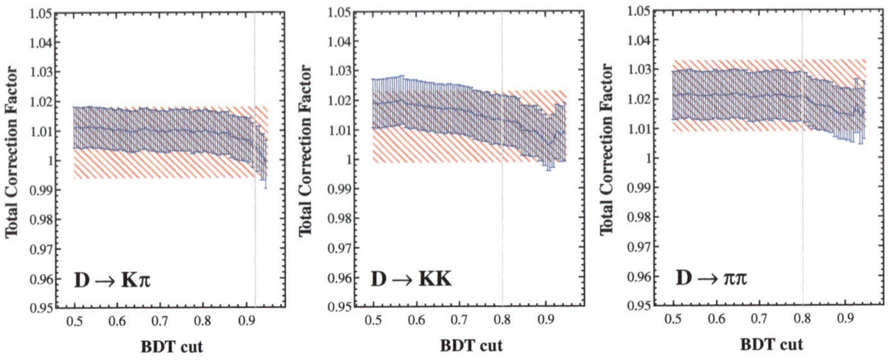

Fig. 5.6 Total correction factors with respect to BDT cut for $B \to [K\pi]_D h$, $B \to [KK]_D h$ and $B \to [\pi\pi]_D h$ signal modes. Corrections are stable over the full BDT range of interest. The *shaded region* indicates the used correction factor with its $\pm 1\sigma$ associated systematic uncertainty

procedure (about 0.02 % absolute) are purely statistical and depend on the large size of the calibration sample only, therefore they could be arbitrary small. A more realistic estimation of the systematic uncertainties associated with PID efficiencies should take into account the specific procedure adopted and in particular its approximations. For example, since it is an approximation that the $\Delta LL^{bach}_{K/\pi}$ distributions are a function of only three variables, the weighting procedure is expected not to reproduce perfectly the $\Delta LL^{bach}_{K/\pi}$ associated with the signal under study and this results in a systematic error on the quoted efficiencies. The size of this systematic uncertainty can be assessed both with MC simulation and data.

PID Systematics: A MC Approach

MC simulated $B \to [hh]_D h$ signal events can be used to give a reasonable estimation of the size of the systematic error associated with $\epsilon^{\text{PID}}_K$and $\epsilon^{\text{PID}}_\pi$, comparing the efficiencies determined on truth matched $B \to [hh]_D h$ MC events with those predicted by the weighting procedure on a MC calibration sample of $D^{*+} \to D^0(K^-\pi^+)\pi^+$ events. The estimation is based on the assumption that:

$$\left[\frac{\Delta\epsilon^{\text{PID}}}{\epsilon^{\text{PID}}_{\text{cal}}}\right]_{\text{DATA}} \approx \left[\frac{\Delta\epsilon^{\text{PID}}}{\epsilon^{\text{PID}}_{\text{cal}}}\right]_{\text{MC}} \quad \text{with} \quad \Delta\epsilon^{\text{PID}} = \epsilon^{\text{PID}}_{\text{sig}} - \epsilon^{\text{PID}}_{\text{cal}}$$

where $\epsilon^{\text{PID}}_{\text{sig}}$is the *true* signal PID efficiency and and $\epsilon^{\text{PID}}_{\text{cal}}$is the corresponding efficiency determined via the calibration procedure. Assuming that the relative uncertainties are the same on MC and data, the absolute difference

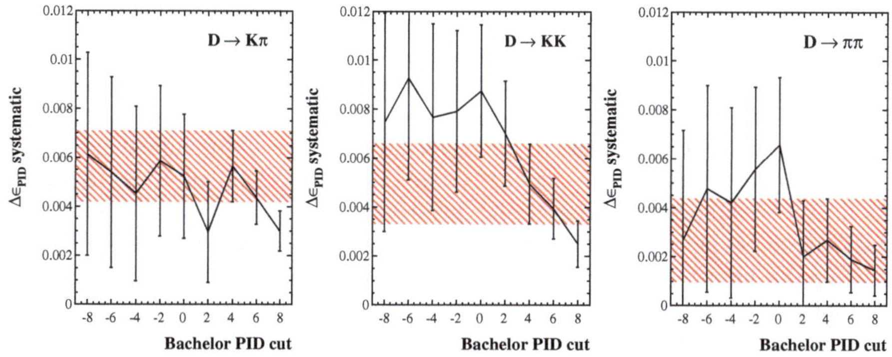

Fig. 5.7 $\Delta\epsilon^{\mathrm{PID}}$ systematic differences between the *true* $\epsilon_{\mathrm{sig}}^{\mathrm{PID}}$ and the calibrated $\epsilon_{\mathrm{cal}}^{\mathrm{PID}}$ for a wide range of $\Delta LL_{K/\pi}^{bach}$cuts. *Left* corrections for $B \to [K\pi]_D\pi$ events. *Center* corrections for $B \to [KK]_D\pi$ events. *Right* corrections for $B \to [\pi\pi]_D\pi$ events. Error bars are propagated from the statistical uncertainties on $\epsilon_{\mathrm{sig}}^{\mathrm{PID}}$ and $\epsilon_{\mathrm{cal}}^{\mathrm{PID}}$, and depend on the size of the MC samples used. The *shaded area* highlights the obtained value for the chosen $\Delta LL_{K/\pi}^{bach} > 4$ cut. $\Delta\epsilon^{\mathrm{PID}}$fluctuates in the 0.2 – 0.7 % range

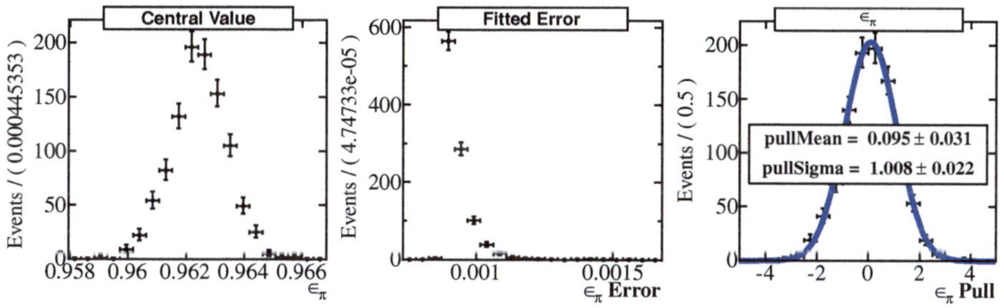

Fig. 5.8 Toy experiments where both the misidentified $B \to D\pi$ shape and $\epsilon_\pi^{\mathrm{PID}}$ are floated independently. *Left* fitted value distribution (to be compared with the generated 3.8 %). *Center* distribution of fitted errors. *Right* resulting pull distributions

$$\left[\Delta\epsilon^{\mathrm{PID}}\right]_{\mathrm{DATA}} = \left[\frac{\Delta\epsilon^{\mathrm{PID}}}{\epsilon_{\mathrm{cal}}^{\mathrm{PID}}}\right]_{\mathrm{MC}} \left[\epsilon_{\mathrm{cal}}^{\mathrm{PID}}\right]_{\mathrm{DATA}}$$

gives a measure of the uncertainty associated with the D^{*+} calibrated PID efficiencies. $[\epsilon_{\mathrm{cal}}^{\mathrm{PID}}]_{\mathrm{DATA}}$ are taken from Tables 3.12–3.15. $\Delta\epsilon^{\mathrm{PID}}$ values are shown in Fig. 5.7 for a wide range of $\Delta LL_{K/\pi}^{bach}$cuts. They suggest differences of 0.2 – 0.7 % over the full range considered, 1 % being a conservative approach to the evaluation of the PID systematic. This value is used for $\epsilon_K^{\mathrm{PID}}$.

Floating the Pion $\epsilon_\pi^{\mathrm{PID}}$ Efficiency on Data

One of the most important parts of the fitting strategy is the treatment of the omnipresent misidentified $B \to D\pi$ component. Two strategies are in principle possible: fix the misidentified $B \to D\pi$ shape and float the the value of the $\pi \to K$ misidentification rate $\epsilon_{\pi}^{\text{PID}}$, or vice versa. The latter solution is preferred in the fit to data as the analysis has an excellent source of external information from the D^* calibration sample to evaluate PID efficiencies on data, as described in Sect. 3.11. The most important parameters of the misidentified $B \to D\pi$ component can thus be floated in the fitter and they do not have to rely on MC simulation. Nevertheless, the fit can be attempted with both the misidentified shape and $\epsilon_{\pi}^{\text{PID}}$ floated.[3] This particular fitter configuration, used as a cross-check, is studied with toy experiments with a procedure similar to that described in Sect. 5.2. Such a fit converges 98 % of the times and no sizable bias can be seen on the extracted value of $\epsilon_{\pi}^{\text{PID}}$, compared to the generated value of 3.8 %. The distributions of the fitted $\epsilon_{\pi}^{\text{PID}}$ parameter and its statistical error are plotted in Fig. 5.8, along with the obtained pull distribution. The error distribution is not perfectly Gaussian and suggests a fitted error of order of 0.1 %. When this fit is performed on data, the pion misidentification rate $\epsilon_{\pi}^{\text{PID}}$ is found to be (0.040 ± 0.001) for the chosen $\Delta LL_{K/\pi}^{bach} = 4$ binary cut. This value is $+0.2$ % away from the the value $\epsilon_{\pi}^{\text{PID}} = 3.8$ % obtained with the PID calibration on the D^* sample (Sect. 3.11), which is introduced in the nominal fit as a fixed parameter. From the comparison of the toy result and the fit to the data, the assumption of a systematic uncertainty of ± 0.2 % on $\epsilon_{\pi}^{\text{PID}}$ seems fully justified. Note that the kaon PID efficiency $\epsilon_{K}^{\text{PID}}$ cannot be floated in the same manner as misidentified $B \to DK$ events contribute negligibly to the FAIL subsample as they are dominated by the most abundant $B \to D\pi$ peak. A conservative systematic of ± 1.0 % is thus assigned to $\epsilon_{K}^{\text{PID}}$.

5.3.3 Propagation Through the Final Fit

Several assumptions are introduced in the fitter as fixed parameters of the model in order to ensure a good stability. Each of these assumptions is assigned a systematic uncertainty and its effect is propagated to the observables of interest. All the systematic effects considered are summarized here:

- **PID efficiencies: $\epsilon_{\pi}^{\text{PID}}$**
  PID efficiencies for correctly reconstructed bachelor kaons and misidentified bachelor pions are introduced in the fitter as fixed numbers. Central values are obtained with the data-driven procedure described in Sect. 3.11, while their associated systematic uncertainties are detailed in Sect. 5.3.2. It is assigned an absolute 0.2 % error on $\epsilon_{\pi}^{\text{PID}}$(central value 3.8 %) and a conservative 1.0 % error on $\epsilon_{K}^{\text{PID}}$(central value 87.2 %).
- **Ratio of the $B \to DK$ and $B \to D\pi$ widths: $R_{\sigma} = \sigma_{B\to DK}/\sigma_{B\to D\pi}$**
  The width of the $B \to DK$ peak is fixed with respect to the width of equivalent $B \to D\pi$ peak. This is discussed in detail in Sect. 5.1.2. The relative ratio is given a systematic of 0.02 (absolute) on a central value of 0.95 for $[K\pi]_D$. The same

[3] although the sigmas of the Crystal Ball function (Sect. 5.1.2) must be fixed to ensure convergence. The tail parameters are allowed to vary, instead.

value is used for the $[\pi K]_D$ subsample. Small possible differences ΔR_σ in resolution for the CP modes $[KK, \pi\pi]_D$ are taken into account with an extra systematic of 0.005 (absolute) on $R_\sigma = -0.005$ and $R_\sigma = +0.005$ for $[\pi\pi]_D$ and $[KK]_D$ respectively with respect to the favoured $[K\pi]_D$.

- **Shape of the misidentified $B \to D\pi$ component in the** PASS **subsample**
  The misidentified $B \to D\pi$ component is described in detail in Sect. 3.9.4 and it is shaped with the sum of two Crystal Ball functions with a common mean for all the D modes. All the parameters of the PDF are floated in the fit, except for the two parameters a and n that define the lower tail of the distribution. These are both fixed from simulation at (3 ± 1).
- **Favoured → suppressed-mode cross feed**
  This type of background, particularly important for the suppressed $B \to [\pi K]_D h$ modes, is rejected with the veto introduced in Sect. 4.5.3. The small residual cross feed rate is estimated from data in Table 4.13. Its ratio with respect to the favoured mode is expected to be $(6 \pm 3) \times 10^{-5}$, where 3×10^{-5} is taken as the systematic variation.
- **$B_s \to [K\pi]_{D^0} K\pi$ background for the suppressed $[\pi K]_D$ mode**
  This background is one of the main uncertainties in the suppressed $B \to [\pi K]_D K$ mode and it is described in Sec. 4.3.2. It is introduced in the fitter with a floating yield parameterised with respect to the favoured $B \to [K\pi]_D \pi$ yield. Its shape is fixed and taken from MC simulation as shown in Fig. 4.18. The distribution is fitted using a bifurcated gaussian with two tail parameters. The position of this principle physics background is varied of $\pm 3\,\text{MeV}/c^2$ on its right edge point and a variation of $\pm 3\,\text{MeV}/c^2$ is assumed on the width on the right-hand side of the fixed MC shape (corresponding to about $\frac{1}{3}$ of the width of the signal peaks and about $\frac{1}{2}$ of the smearing introduced in MC simulated events, see Sect. 3.9), which leaks under the signal component. These values are conservative variations if compared to the error on resolution of the peaks.
- **Production and detection asymmetries:** A_{instr}
  Charge asymmetries are introduced in Sect. 5.1.4. Along with the weak CP-violation physical asymmetries, other sources of charge asymmetries can be introduced, e.g the production asymmetry $A_{\text{prod}}$and the kaon/pion detection asymmetries $A_K^{\text{det}}/A_\pi^{\text{det}}$. The sum of all these effects is referred to as A_{instr} and it includes the following systematics:
  - 0.7 % (absolute) on central values of 0.0 % for A_π^{det}
  - 0.7 % (absolute) on central values of −0.5 % for A_K^{det}
  - 0.7 % (absolute) on central values of −0.8 % for A_{prod}
- **Shape of the misidentified $B \to DK$ component in the** FAIL **subsample**
  This component takes into account $B \to DK$ signal events which fail the $\Delta LL_{K/\pi}^{bach}$ requirement and it is included in the fit as a PDF (sum of two Crystal Ball functions) with all the parameters fixed to MC simulation. A $\pm 5\,\text{MeV}/c^2$ absolute

shift on the MC mean and a $\pm 5\,\mathrm{MeV}/c^2$ absolute variation on the MC widths are allowed as systematic uncertainties. These values are conservative variations if compared to the error on resolution of the peaks.

- **Cabibbo suppression factor in the $\Lambda_b \to [Kp\pi]_{\Lambda_c} h$ background:** $R_{K/\pi}(\Lambda_b)$
  In the $[KK]_D$ mode, two specific floating components are introduced to take into account $\Lambda_b \to [Kp\pi]_{\Lambda_c} h$ decays where the daughter proton from the Λ_c has been reconstructed as a kaon (see Sect. 4.3.2). Both $\Lambda_b \to \Lambda_c \pi$ and $\Lambda_b \to \Lambda_c K$ (this decay is currently unobserved yet) contributions are included in the fit, although the small $\Lambda_b \to \Lambda_c K$ component is assumed to have the identical shape as its pion Cabibbo-favoured counterpart under the kaon mass hypothesis. The $\Lambda_b \to \Lambda_c K / \Lambda_b \to \Lambda_c \pi$ ratio is fixed at 6.0 % and a systematic uncertainty of 1.5 % is associated on this estimated value.
- **Multiplicative corrections on**
  The correction factors on $R^{hh}_{K/\pi}$ are described in detail in Sect. 5.3.1. Every multiplicative factor carries a systematic uncertainty of 0.011 on combined correction factors in the range of 1.006 – 1.021 as shown in Fig. 5.6.
- **Partially reconstructed background PDF**
  Partially reconstructed backgrounds are shaped with a non-parametric PDF shaped on simulated events as described in Sect. 4.3.2. This shape is applied to all D modes, although each yield is allowed to vary independently. A systematic is assigned on this shape by modifying it accordingly allowing an extra exponential tail in the component in order to increase the slope of the tail leaking in the signal window. The size of this extra exponential is varied such that it contains up to 10 % of the total partially reconstructed PDF.

To evaluate the effect of these systematic uncertainties, the fit is rerun many times varying each of the systematic constants (the fixed parameters) according to a Gaussian whose width is the assigned uncertainty. For each parameter of the fit, a histogram is filled with as many entries as the repeated fits that have been performed (about 200 times for each systematic source). Histograms are plotted in Appendix C. The width σ of a fitted Gaussian, or the resulting spread (RMS) of the histogram if the distribution is too skewed, is taken as the systematic uncertainty on that quantity. All contributions are summarised in Table 5.2. Correlations between the uncertainties are found negligible so the total systematic uncertainty is just the sum in quadrature of each individual source. For the branching fraction ratios $R^{hh}_{K/\pi}$ in the favoured and CP modes, the uncertainties on the PID efficiencies, the multiplicative factors and the misidentified $B \to D\pi$ shape dominate. These sources also contribute in the suppressed modes on $R^{\pm}_{K}$ and $R^{\pm}_{\pi}$, though the assumed shape of the $B_s \to [K\pi]_{D^0} K\pi$ background is one of the largest source of systematic uncertainty in the $B \to [\pi K]_D K$ case. As expected, for the CP asymmetries the instrumentation asymmetries A_{instr} are the largest source of uncertainty, since they enter directly in their computation.

Table 5.2 Systematic uncertainties on the fitted *CP* observables

$\times 10^{-2}$	$R^{K\pi}_{K/\pi}$	$R^{KK}_{K/\pi}$	$R^{\pi\pi}_{K/\pi}$	$A^{K\pi}_{\pi}$	$A^{K\pi}_{K}$	A^{KK}_{K}	$A^{\pi\pi}_{K}$
$\epsilon^{\rm PID}_{K}$	0.09	0.09	0.10	0.00	0.00	0.03	0.14
$\epsilon^{\rm PID}_{\pi}$	0.10	0.10	0.09	0.00	0.02	0.16	0.13
R_σ	0.05	0.03	0.02	0.00	0.01	0.09	0.06
ΔR_σ	0.00	0.01	0.01	0.00	0.00	0.01	0.01
misID $B \to D\pi$	0.06	0.07	0.04	0.10	0.41	0.05	0.05
Cross feed	0.00	0.00	0.00	0.00	0.00	0.00	0.00
$A_{\rm instr}$	0.00	0.00	0.00	0.94	1.69	0.95	0.90
misID $B \to DK$	0.02	0.02	0.01	0.00	0.01	0.05	0.20
$R_{K/\pi}(\Lambda_b)$	0.00	0.04	0.00	0.00	0.00	0.05	0.00
Mult. Corr.	0.08	0.08	0.08	0.00	0.00	0.00	0.00
$B_s \to [K\pi]_{D^0} K\pi$	0.00	0.00	0.00	0.00	0.01	0.01	0.01
Part reco bkg	0.05	0.04	0.03	0.01	0.00	0.06	0.08
Total$\times 10^{-2}$ (absolute)	0.18	0.18	0.17	0.95	1.74	0.97	0.95
Relative (%)	(2.3)%	(2.3)%	(2.1)%	–	–	(6.6)%	(7.0)%

$\times 10^{-2}$	A^{KK}_{π}	$A^{\pi\pi}_{\pi}$	R^{-}_{K}	R^{+}_{K}	R^{-}_{π}	R^{+}_{π}
$\epsilon^{\rm PID}_{K}$	0.01	0.01	0.00	0.01	0.00	1e-3
$\epsilon^{\rm PID}_{\pi}$	0.00	0.00	0.02	0.04	1e-3	1e-3
R_σ	0.00	0.00	0.01	0.01	0.00	0.00
ΔR_σ	0.00	0.00	0.00	0.00	0.00	0.00
misID $B \to D\pi$	0.66	0.03	0.01	0.01	0.00	0.00
Cross feed	0.00	0.00	0.00	0.00	3e-3	2e-3
$A_{\rm instr}$	0.95	0.99	0.01	0.01	7e-3	7e-3
misID $B \to DK$	0.01	0.01	0.01	0.01	0.00	1e-3
$R_{K/\pi}(\Lambda_b)$	0.00	0.00	0.00	0.00	0.00	0.00
Mult. Corr.	0.00	0.00	0.00	0.00	0.00	0.00
$B_s \to [K\pi]_{D^0} K\pi$	0.00	0.00	0.03	0.04	1e-3	0.00
Part reco bkg	0.00	0.03	0.01	0.02	2e-3	1e-3
Total$\times 10^{-2}$ (absolute)	1.16	0.99	0.04	0.07	8e-3	7e-3
Relative (%)	–	–	(5.5)%	(3.0)%	(1.7)%	(2.0)%

References

1. W. Verkerke, D. Kirkby, The RooFit toolkit for data modeling, arXiv:physics/0306116
2. LHCb collaboration, R. Aaij, and others, Measurements of the branching fractions and CP asymmetries of $B^+ \to J/\psi\pi^+$ and $B^+ \to \psi(2S)\pi^+$ decays. Phys. Rev. D85 091105, arXiv:1203.3592 (2012)
3. R. Aaji and others, Measurement of the production asymmetry in 7 TeV pp collisions. Phys. Lett. B **713**(3), p. 186 (2012)

Chapter 6
Results

This chapter summarises the numerical results of the simultaneous fit to data. Section 6.1 presents the output of the fit with some internal cross-checks and discussion on the fitted parameters. The number of $B \to [hh]_D\pi$ events is a floating parameter in the fit itself and it is compared with MC expectations in Sect. 6.2. Section 6.3 presents a summary of the 13 fitted *CP* observables, along with their combination and a likelihood scan to compute the significance of this measurement (Sect. 6.4). Once errors are propagated to the final result and *CP* violation effects are considered simultaneously, it is claimed that direct *CP* violation is observed for the first time in $B^\pm \to DK^\pm$ decays with a total significance of 5.8σ. The chapter is then concluded with a comparison with the existing results (Sect. 6.5) and a preliminary study on how to fit the CKM angle γ with the measured physics observables (Sect. 6.6).

6.1 Numerical Fit Results

The full simultaneous fit, applied to all the $[hh]_D$ modes, is shown in Figs. 6.1, 6.2, 6.3, 6.4 with the fit results printed below for any floated parameter, excluding the relatively uninteresting combinatorial and partially reconstructed background yields. The final $B \to [hh]_D h$ signal yields, after summing the events that pass and fail the bachelor $\Delta LL^{bach}_{K/\pi}$ PID cut and split by charge, are shown in Table 6.1. The $B \to [hh]_D\pi$ mass resolutions are floated independently and found to be $\sigma_{B\to[KK]_D\pi} = 14.1 \pm 0.1\,\text{MeV}/c^2$, $\sigma_{B\to[K\pi]_D\pi} = 14.2 \pm 0.1\,\text{MeV}/c^2$ and $\sigma_{B\to[\pi\pi]_D\pi} = 14.2 \pm 0.2\,\text{MeV}/c^2$. These values are compatible as expected from the D mass constraint procedure (Sect. 3.9.2), which removes most of the dependence on the particular $[hh]_D$ submode considered. As explained in Sect. 5.1.2, the $B \to [hh]_D K$ widths are fixed relative to these values. The fitted common tail parameters of Eq. 3.1 are $\alpha_L = 0.115 \pm 0.003$ and $\alpha_R = 0.083 \pm 0.002$.

P. Gandini, *Observation of CP Violation in $B^\pm \to DK^\pm$ Decays*,
Springer Theses, DOI: 10.1007/978-3-319-01029-8_6,

```
RooFitResult: minimized FCN value: -232483, estimated distance to
minimum: 0.00308009 coviarance matrix quality: Full, accurate
covariance matrix

    Floating Parameter    InitialValue    FinalValue +/-  Error
  --------------------  ------------  --------------------------
         A_CPP_b2dk_d2kk   2.0000e-01    1.4798e-01 +/-  3.69e-02
       A_CPP_b2dk_d2pipi   2.0000e-01    1.3506e-01 +/-  6.61e-02
        A_CPP_b2dpi_d2kk  -0.0000e+00   -1.9924e-02 +/-  9.11e-03
      A_CPP_b2dpi_d2pipi  -0.0000e+00   -8.7475e-04 +/-  1.65e-02
        A_FAV_b2dk_d2kpi  -8.0000e-03    4.4024e-03 +/-  1.45e-02
       A_FAV_b2dpi_d2kpi   3.0000e-03   -9.3880e-05 +/-  3.57e-03
        R_dk_vs_dpi_d2kk   7.6000e-02    7.7259e-02 +/-  3.02e-03
       R_dk_vs_dpi_d2kpi   7.6000e-02    7.7418e-02 +/-  1.24e-03
      R_dk_vs_dpi_d2pipi   7.6000e-02    8.0284e-02 +/-  5.58e-03
      R_minus_b2dk_d2pik   3.0000e-02    7.3110e-03 +/-  2.27e-03
     R_minus_b2dpi_d2pik   3.0000e-02    4.6857e-03 +/-  3.78e-04
       R_plus_b2dk_d2pik   3.0000e-02    2.3178e-02 +/-  3.38e-03
      R_plus_b2dpi_d2pik   3.0000e-02    3.5163e-03 +/-  3.34e-04
        alphaL_dpi_d2kpi   1.0000e-01    1.1467e-01 +/-  3.35e-03
        alphaR_dpi_d2kpi   7.0000e-02    8.2806e-02 +/-  2.27e-03
               comb_coef  -2.0000e-02   -5.9124e-01 +/-  5.35e-02
                  mean_B   5.2790e+03    5.2786e+03 +/-  6.50e-02
       n_b2dpi_d2kk_both   1.3000e+04    1.3343e+04 +/-  1.32e+02
      n_b2dpi_d2kpi_both   9.5000e+04    8.1541e+04 +/-  3.01e+02
     n_b2dpi_d2pipi_both   4.5000e+03    3.9419e+03 +/-  6.56e+01
          sigma_dpi_d2kk   1.4300e+01    1.4056e+01 +/-  1.36e-01
         sigma_dpi_d2kpi   1.4300e+01    1.4242e+01 +/-  7.28e-02
        sigma_dpi_d2pipi   1.4300e+01    1.4145e+01 +/-  2.23e-01
```

Crosschecks with Only `L0Hadron_TOS` *Triggers*

More than 35 % of events come from `L0Global_TIS`triggers which are accepted by the selection as described in Sect. 3.6 (although overlap is expected with TOS events). TIS events carry an uncertainty due to the inclusive, so unknown, nature of the triggered object. The default result has been thus compared with that once `L0Global_TIS`triggers are removed, in order to double-check any possible bias in the measured physics quantities. This cross-check is shown in Table 6.2 and it shows that there is no evidence of a consistent bias in the *CP* observables.

6.2 Yield Comparison

The $[K\pi]_D$, $[KK]_D$ and $[\pi\pi]_D$ slices are fitted simultaneously but the number of $B \to D\pi$ candidates is floated independently for each D submode and it is thus a free parameter of the extended ML fit. The fitted $n_{B\to[K\pi]_D\pi}$, $n_{B\to[KK]_D\pi}$

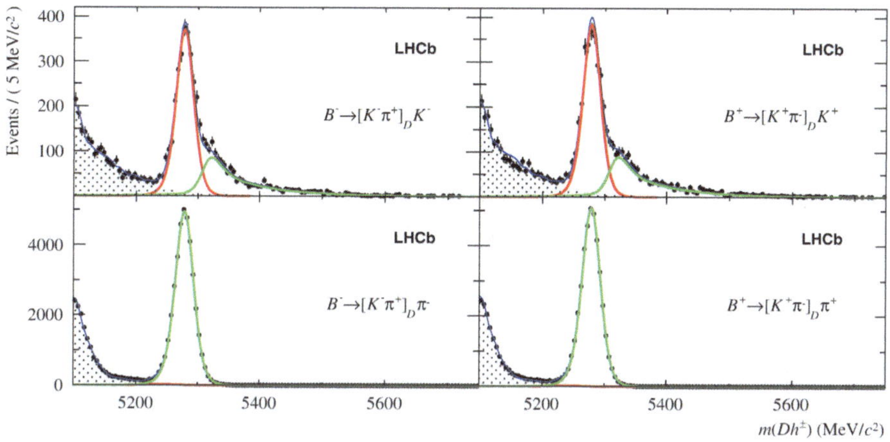

Fig. 6.1 Invariant mass distributions of selected favoured $B \to [K\pi]_D h$ candidates, split by charge. The *left* plots are B^- candidates, B^+ are on the *right*. In the *top* plots, the bachelor track passes the $\mathrm{DLL}_{K\pi} > 4$ cut (PASS slice) and the B candidates are reconstructed assigning this track the kaon mass. The remaining events are placed in the sample displayed on the *bottom row* (FAIL slice) and are reconstructed with a pion mass hypothesis. The *red curve* represents the $B \to DK$ events, the *green curve* is $B \to D\pi$. Partially reconstructed events are represented by the shaded *gray* area

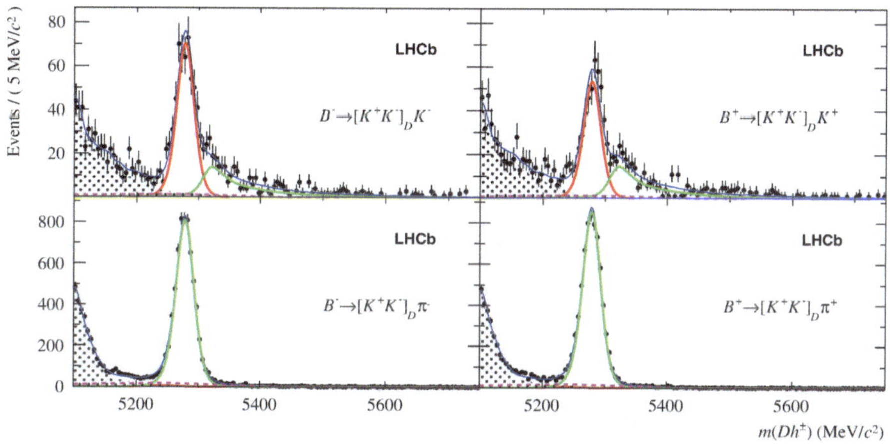

Fig. 6.2 Invariant mass distributions of selected $B \to [KK]_D h$ candidates. See the caption of Fig. 6.1 for a full description. The small contribution from $\Lambda_b \to \Lambda_c h$ decays is indicated by the *dashed purple line*

and $n_{B\to[\pi\pi]_D\pi}$ can be used to check the internal consistency of the dataset used. The fitted signal yields are compared with MC expectations in Table 6.3, where expectations are obtained via:

$$n_{B\to[hh]_Dh} = \mathcal{L}_{\mathbf{int}} \cdot \sigma_{(pp\to b\bar{b}X)} \cdot f_{\mathbf{had}} \cdot \mathcal{B}_{B\to[hh]_Dh} \cdot \varepsilon_{\mathrm{eff}}^{\mathrm{Geom}} \cdot \varepsilon_{\mathrm{eff}}^{\mathrm{Sel}} \cdot \varepsilon_{\mathrm{eff}}^{\mathrm{PID_{dau}}} \cdot f_{\mathrm{corr}}$$

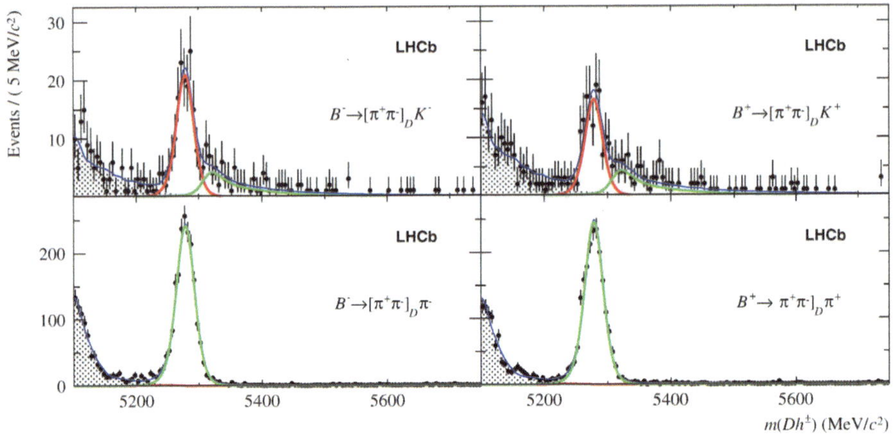

Fig. 6.3 Invariant mass distributions of selected $B \to [\pi\pi]_D h$ candidates. See the caption of Fig. 6.1 for a full description

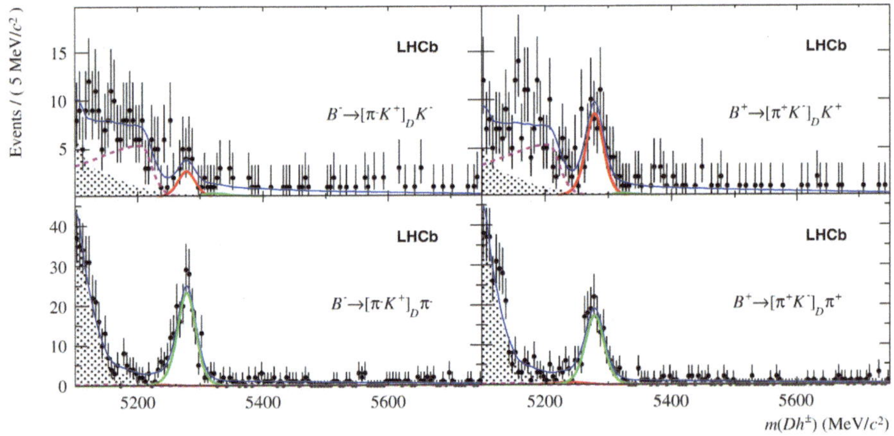

Fig. 6.4 Invariant mass distributions of selected suppressed $B \to [\pi K]_D h$ candidates. See the caption of Fig. 6.1 for a full description. The dashed line here represents the partially reconstructed, but Cabibbo favoured, $B_s^0 \to \overline{D}^0 K^-\pi^+$ and $\overline{B}_s^0 \to D^0 K^+\pi^-$ decays where the pions are lost. The pollution from favoured mode cross feed is drawn, but is too small to be seen

as detailed in Sect. 3.8. The multiplicative factors that are common with all the $B \to [hh]_D h$ modes are highlighted here in gray. The yields found on data are consistently lower than MC estimations by ~30 %. This factor acts as an overall multiplicative effect which might depend either on the measured of the $\sigma_{(pp\to b\bar{b}X)}$ production cross-section (which has a relative error of about 20 %) or an over-optimistic reconstruction efficiency of the B candidates in the MC simulation of the LHCb detector. Nevertheless, the fitted yields are still compatible at the level of 1.5σ, once all errors are taken into account.

Table 6.1 Fitted yields for each $B \to [hh]_D h$ mode, split by charge

$B^{\pm}$ mode	D mode	B^-	B^+	Total
$D\pi^{\pm}$	$K^{\pm}\pi^{\mp}$	40767 ± 310	40774 ± 310	81541 ± 438
	$K^{\pm}K^{\mp}$	6539 ± 129	6804 ± 135	13343 ± 187
	$\pi^{\pm}\pi^{\mp}$	1969 ± 69	1973 ± 69	3942 ± 98
	$\pi^{\pm}K^{\mp}$	191 ± 16	143 ± 14	334 ± 21
$DK^{\pm}$	$K^{\pm}\pi^{\mp}$	3170 ± 83	3142 ± 83	6312 ± 117
	$K^{\pm}K^{\mp}$	592 ± 40	439 ± 30	1031 ± 50
	$\pi^{\pm}\pi^{\mp}$	180 ± 22	137 ± 16	317 ± 27
	$\pi^{\pm}K^{\mp}$	23 ± 7	73 ± 11	96 ± 13

Table 6.2 *CP* observables once TIS triggers are removed compared to the trigger strategy adopted in the default trigger selection (Sect. 3.6)

	TOS *or* TIS	TOS only	Shift
A_K^{KK}	1.4798e-01 $\pm$ 3.69e-02	1.0048e-01	-1.2σ
$A_K^{\pi\pi}$	1.3506e-01 $\pm$ 6.61e-02	1.9024e-01	$+1.0\sigma$
A_π^{KK}	$-$1.9924e-02 $\pm$ 9.11e-03	$-$1.7835e-02	$+0.3\sigma$
$A_\pi^{\pi\pi}$	$-$8.7475e-04 $\pm$ 1.65e-02	$-$1.3564e-02	-0.8σ
$A_K^{K\pi}$	4.4024e-03 $\pm$ 1.45e-02	3.4188e-04	-0.3σ
$A_\pi^{K\pi}$	$-$9.3880e-05 $\pm$ 3.57e-03	$-$4.0892e-03	-1.2σ
$R_{K/\pi}^{KK}$	7.7259e-02 $\pm$ 3.02e-03	7.7400e-02	-0.0σ
$R_{K/\pi}^{K\pi}$	7.7418e-02 $\pm$ 1.24e-03	7.6677e-02	-0.6σ
$R_{K/\pi}^{\pi\pi}$	8.0284e-02 $\pm$ 5.58e-03	7.4845e-02	-0.9σ
R_K^-	7.3110e-03 $\pm$ 2.27e-03	7.4145e-03	$+0.4\sigma$
R_π^-	4.6857e-03 $\pm$ 3.78e-04	4.6843e-03	0.0σ
R_K^+	2.3178e-02 $\pm$ 3.38e-03	2.4553e-02	$+0.4\sigma$
R_π^+	3.5163e-03 $\pm$ 3.34e-04	3.8195e-03	$+0.8\sigma$

As an additional cross-check, it is possible to compare directly the ratio of the $B \to [K\pi]_D\pi$, $B \to [KK]_D\pi$ and $B \to [\pi\pi]_D\pi$ yields with the ratio of their branching fractions. In this way, the common multiplicative factors introduced before cancel out completely (although mode-dependent correction factors must be still considered). The relative ratio of the fitted yields is presented in Table 6.4. The branching fraction ratios are corrected as follows:

$$\frac{n_{B\to[h'h']_D\pi}}{n_{B\to[hh]_D\pi}} = \frac{\mathcal{B}_{D\to h'h'}}{\mathcal{B}_{D\to hh}} \cdot \frac{\varepsilon_{\text{eff}}^{\text{Geom}'}}{\varepsilon_{\text{eff}}^{\text{Geom}}} \cdot \frac{\varepsilon_{\text{eff}}^{\text{Sel}'}}{\varepsilon_{\text{eff}}^{\text{Sel}}} \cdot \frac{\varepsilon_{\text{eff}}^{\text{PID}'_{\text{dau}}}}{\varepsilon_{\text{eff}}^{\text{PID}_{\text{dau}}}} \cdot \frac{f'_{\text{corr}}}{f_{\text{corr}}}$$

All ratios are compatible between 1σ from the expected value.

Table 6.3 Comparison of the $n_{B\to[hh]_D h}$ yields fitted on data with MC expectations

$B^{\pm}$ mode	D mode	Fitted yield	Expected yield	Shift
$D\pi^{\pm}$	$K^{\pm}\pi^{\mp}$	81541 ± 438	114505 ± 22084	-1.49σ
	$K^{\pm}K^{\mp}$	13343 ± 187	18805 ± 3648	-1.50σ
	$\pi^{\pm}\pi^{\mp}$	3942 ± 98	4921 ± 1126	-0.87σ
$DK^{\pm}$	$K^{\pm}\pi^{\mp}$	6312 ± 117	8757 ± 1841	-1.33σ
	$K^{\pm}K^{\mp}$	1031 ± 50	1448 ± 306	-1.34σ
	$\pi^{\pm}\pi^{\mp}$	317 ± 27	382 ± 93	-0.67σ

Fitted yield are taken from Table 6.1. MC expectations are taken from Table 3.7. The fitted yields are all consistently lower compared with the fitted ones, but still compatible at the level of 1.5σ maximum, once errors on MC estimations are taken into account

Table 6.4 Relative ratio of $n_{B\to[hh]_D\pi}$ for the yields as fitted in data and the MC expectations

	Measured ratio	MC estimation	Shift
$\frac{n_{B\to[KK]_D\pi}}{n_{B\to[K\pi]_D\pi}}$	0.1636 ± 0.0025	0.1642 ± 0.0089	-0.06σ
$\frac{n_{B\to[\pi\pi]_D\pi}}{n_{B\to[K\pi]_D\pi}}$	0.0483 ± 0.0012	0.0430 ± 0.0057	$+0.92\sigma$
$\frac{n_{B\to[\pi\pi]_D\pi}}{n_{B\to[KK]_D\pi}}$	0.2954 ± 0.0084	0.2617 ± 0.0352	$+0.93\sigma$

All multiplicative factors like the production cross-section and the hadronisation factor cancel out completely

6.3 The *CP* Observables

The fitted *CP* observables are summarized here with their statistical uncertainties and assigned systematic uncertainties

$$\frac{\Gamma(B^- \to [K^-\pi^+]_D K^-) + \Gamma(B^+ \to [K^+\pi^-]_D K^+)}{\Gamma(B^- \to [K^-\pi^+]_D \pi^-) + \Gamma(B^+ \to [K^+\pi^-]_D \pi^+)} = R^{K\pi}_{K/\pi} = 0.0774 \pm 0.0012 \pm 0.0018$$

$$\frac{\Gamma(B^- \to [K^-K^+]_D K^-) + \Gamma(B^+ \to [K^+K^-]_D K^+)}{\Gamma(B^- \to [K^-K^+]_D \pi^-) + \Gamma(B^+ \to [K^+K^-]_D \pi^+)} = R^{KK}_{K/\pi} = 0.0773 \pm 0.0030 \pm 0.0018$$

$$\frac{\Gamma(B^- \to [\pi^-\pi^+]_D K^-) + \Gamma(B^+ \to [\pi^+\pi^-]_D K^+)}{\Gamma(B^- \to [\pi^-\pi^+]_D \pi^-) + \Gamma(B^+ \to [\pi^+\pi^-]_D \pi^+)} = R^{\pi\pi}_{K/\pi} = 0.0803 \pm 0.0056 \pm 0.0017$$

$$\frac{\Gamma(B^- \to [K^-\pi^+]_D \pi^-) - \Gamma(B^+ \to [K^+\pi^-]_D \pi^+)}{\Gamma(B^- \to [K^-\pi^+]_D \pi^-) + \Gamma(B^+ \to [K^+\pi^-]_D \pi^+)} = A^{K\pi}_{\pi} = -0.0001 \pm 0.0036 \pm 0.0095$$

$$\frac{\Gamma(B^- \to [K^-\pi^+]_D K^-) - \Gamma(B^+ \to [K^+\pi^-]_D K^+)}{\Gamma(B^- \to [K^-\pi^+]_D K^-) + \Gamma(B^+ \to [K^+\pi^-]_D K^+)} = A^{K\pi}_{K} = 0.0044 \pm 0.0144 \pm 0.0174$$

$$\frac{\Gamma(B^- \to [K^-K^+]_D K^-) - \Gamma(B^+ \to [K^+K^-]_D K^+)}{\Gamma(B^- \to [K^-K^+]_D K^-) + \Gamma(B^+ \to [K^+K^-]_D K^+)} = A^{KK}_{K} = 0.148 \pm 0.037 \pm 0.010$$

$$\frac{\Gamma(B^- \to [\pi^-\pi^+]_D K^-) - \Gamma(B^+ \to [\pi^+\pi^-]_D K^+)}{\Gamma(B^- \to [\pi^-\pi^+]_D K^-) + \Gamma(B^+ \to [\pi^+\pi^-]_D K^+)} = A^{\pi\pi}_{K} = 0.135 \pm 0.066 \pm 0.010$$

$$\frac{\Gamma(B^- \to [K^-K^+]_D \pi^-) - \Gamma(B^+ \to [K^+K^-]_D \pi^+)}{\Gamma(B^- \to [K^-K^+]_D \pi^-) + \Gamma(B^+ \to [K^+K^-]_D \pi^+)} = A^{KK}_{\pi} = -0.020 \pm 0.009 \pm 0.012$$

$$\frac{\Gamma(B^- \to [\pi^-\pi^+]_D \pi^-) - \Gamma(B^+ \to [\pi^+\pi^-]_D \pi^+)}{\Gamma(B^- \to [\pi^-\pi^+]_D \pi^-) + \Gamma(B^+ \to [\pi^+\pi^-]_D \pi^+)} = A^{\pi\pi}_{\pi} = -0.001 \pm 0.017 \pm 0.010$$

$$\frac{\Gamma(B^- \to [\pi^- K^+]_D K^-)}{\Gamma(B^- \to [K^-\pi^+]_D K^-)} = R_K^- = \quad 0.0073 \pm 0.0023 \pm 0.0004$$
$$\frac{\Gamma(B^+ \to [\pi^+ K^-]_D K^+)}{\Gamma(B^+ \to [K^+\pi^-]_D K^+)} = R_K^+ = \quad 0.0232 \pm 0.0034 \pm 0.0007$$
$$\frac{\Gamma(B^- \to [\pi^- K^+]_D \pi^-)}{\Gamma(B^- \to [K^-\pi^+]_D \pi^-)} = R_\pi^- = \quad 0.00469 \pm 0.00038 \pm 0.00008$$
$$\frac{\Gamma(B^+ \to [\pi^+ K^-]_D \pi^+)}{\Gamma(B^+ \to [K^+\pi^-]_D \pi^+)} = R_\pi^+ = \quad 0.00352 \pm 0.00033 \pm 0.00007$$

The favored $B \to D\pi$ and $B \to DK$ asymmetries are found consistent with zero. This provides a useful cross-check on the assumptions introduced in Sect. 5.1.4 on the instrumentation asymmetries. The $[KK]_D$ and $[\pi\pi]_D$ asymmetries and ratios are fitted separately as an internal cross-check (they should be in principle the same). From these measurements, the following quantities can be deduced:

$$R_{CP+} \approx \frac{< R_{K/\pi}^{KK}, R_{K/\pi}^{\pi\pi} >}{R_{K/\pi}^{K\pi}} = \quad 1.007 \pm 0.038 \pm 0.012 \tag{6.1}$$

$$\mathbf{A}_{CP+} =< A_K^{KK}, A_K^{\pi\pi} > = \quad 0.145 \pm 0.032 \pm 0.010 \tag{6.2}$$

$$R_{\mathbf{ADS}(K)} = \frac{R_K^- + R_K^+}{2} = \quad 0.0152 \pm 0.0020 \pm 0.0004 \tag{6.3}$$

$$A_{\mathbf{ADS}(K)} = \frac{R_K^- - R_K^+}{R_K^- + R_K^+} = -0.52 \pm 0.15 \pm 0.02 \tag{6.4}$$

$$R_{\mathbf{ADS}(\pi)} = \frac{R_\pi^- + R_\pi^+}{2} = \quad 0.00410 \pm 0.00025 \pm 0.00005 \tag{6.5}$$

$$A_{\mathbf{ADS}(\pi)} = \frac{R_\pi^- - R_\pi^+}{R_\pi^- + R_\pi^+} = \quad 0.143 \pm 0.062 \pm 0.011 \tag{6.6}$$

where the correlations between systematic uncertainties are taken into account in the combination and angled brackets indicate weighted averages.
The above definition of R_{CP+} is only approximate and is used for experimental convenience. It assumes the absence of *CP* violation in $B \to D\pi$ and the favoured $B \to DK$ modes. The exact definition of R_{CP+} is

$$\frac{\Gamma(B^- \to D_{CP+}K^-) + \Gamma(B^+ \to D_{CP+}K^+)}{\Gamma(B^- \to D^0 K^-)} \tag{6.7}$$

so an additional 1 % systematic uncertainty accounts for the approximation (see Sect. 1.6.3). This systematic uncertainty is obtained as the maximum deviation of the approximate formula from the exact one, once the region of interest of the $(r_B^K, \delta_B^K, \gamma)$ parameters is considered.

6.4 Statistical Significance of the Measurements

The statistical significance of each result is obtained comparing the logarithmic likelihood difference between the minimum value and its value in the null hypothesis $\sqrt{-2\Delta LL}$. For the significance of the suppressed peak, the null hypothesis is defined as the absence of signal $n_{B\to[\pi K]_D K} = 0$, while *CP* violation is tested against $A_{B\to[hh]_D K} = 0$. The suppressed $B \to [\pi K]_D K$ mode is observed for the first time with $\sim 10\sigma$ statistical significance when comparing the maximum likelihood to that of the null hypothesis. This mode displays evidence (4.0σ) of a large negative asymmetry, consistent with the LHCb preliminary first result [1] with $342.7\,\text{pb}^{-1}$. The suppressed $B \to [\pi K]_D \pi$ mode exhibits a hint of a positive asymmetry with 2.4σ significance. The $[KK]_D$ and $[\pi\pi]_D$ *CP* modes both show positive asymmetries of around 14 % (consistent between each other) and the significance of this asymmetry is 4.5σ. Assuming the *CP*-violating effects in the *CP* and **ADS** modes are due to the same phenomenon (namely the interference of $b \to c\bar{u}s$ and $b \to u\bar{c}s$ transitions), the combination of the *CP* violating effects in $[KK, \pi\pi, K\pi, \pi K]_D$ has a statistical significance of 6.0σ. This log-likelihood difference is diluted by the non-negligible systematic uncertainties in $\mathbf{A}_{CP+}$ and $A_{\mathbf{ADS}(K)}$ which are dominated by the instrumentation asymmetries. The systematic dilution is realized convolving the likelihood curve with a Gaussian whose width represents the size of the systematic uncertainty associated to the quantity considered. The likelihood scan of $A_{\mathbf{ADS}(K)}$ + $\mathbf{A}_{CP+}$ is shown in Fig. 6.5. In conclusion, with a total significance of 5.8σ, direct *CP* violation in $B \to DK$ decays is observed.

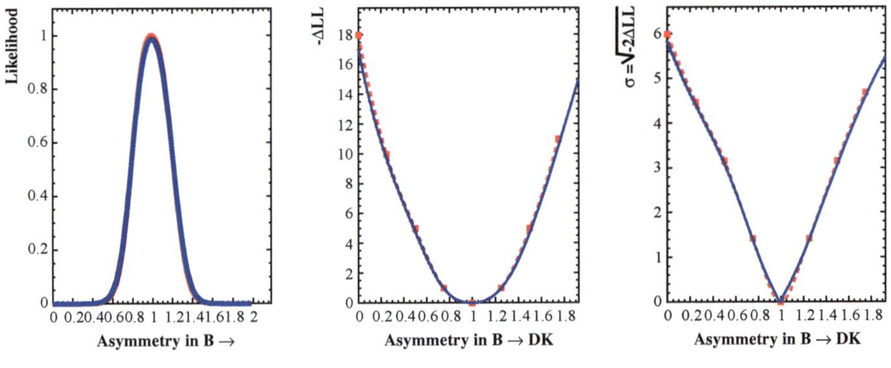

Fig. 6.5 A likelihood scan of the percentage of the central value of the *CP*-violating observables $\mathbf{A}_{CP+}$ and $A_{\mathbf{ADS}(K)}$. This is defined as the distance from the minimum in the ($\mathbf{A}_{CP+}$,$A_{\mathbf{ADS}(K)}$) plane with respect to the origin, corresponding to the absence of *CP* violation. The *dashed curve* is the pure statistical error, the *solid curve* shows the dilution of the likelihood due to an assigned systematic with a Gaussian approximation. The minimum is found to be 5.8σ away from the null hypothesis

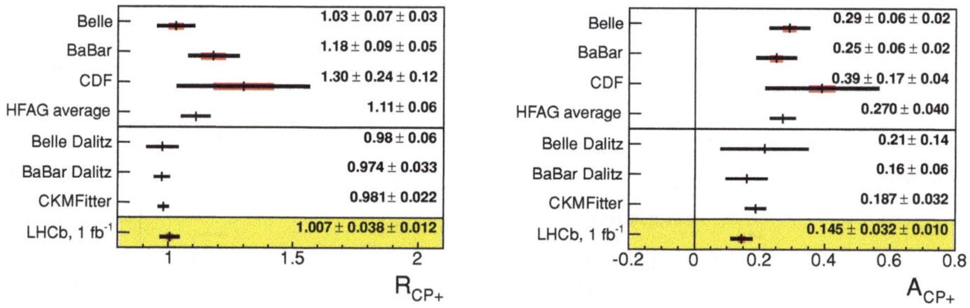

Fig. 6.6 Comparison plot of the measured R_{CP+} and $\mathbf{A}_{CP+}$ observables. The first rows refer to the direct measurements by BaBar [3], Belle [4, 5] and CDF [6]. The second part of the plot refers to the same quantities, once those are calculated from the best values of r_B^K, δ_B^K and γ as extracted by BaBar [7], Belle [8] and CKMfitter [9]

6.5 Comparison with Existing Measurements

This section presents a discussion of the results presented in Sect. 6.3, including a detailed comparison with the existing measurements.

- **Favoured modes:** $A_{\pi}^{K\pi}$, $A_{K}^{K\pi}$ **and** $R_{K/\pi}^{K\pi}$
  The charge asymmetries for both favoured $B \to [K\pi]_D h$ decays are floating parameters in the fit and they are found to be consistent with 0. This provides a good cross-check to the fitter itself and to the size of the systematical uncertainty introduced for the production and detection asymmetry in Sect. 5.1.4. The measured ratio of branching fractions $R_{K/\pi}^{K\pi} = 0.0774 \pm 0.0012 \pm 0.0018$ is consistent with the world average 0.076 ± 0.006 [2].
- ***CP*** **modes**: R_{CP+} **and** $\mathbf{A}_{CP+}$
  The charge asymmetries for $B \to [KK]_D\pi$ and $B \to [\pi\pi]_D\pi$ decays are consistent with 0, as expected. $B \to [KK]_D K$ and $B \to [\pi\pi]_D K$ observables are combined to extract R_{CP+} and $\mathbf{A}_{CP+}$ (Eqs. 6.1 and 6.2), which are then compared with the current measurement from BaBar [3], Belle [4, 5] and CDF [6] in Fig. 6.6. Both R_{CP+} and $\mathbf{A}_{CP+}$ are found to be lower than what observed by the other experiments (although statistically compatible). However, a perfect agreement is observed once those quantities are calculated from the best values of the fundamental quantities r_B^K, δ_B^K and γ from BaBar's and Belle's Dalitz analyses [7, 8] (see Sect. 6.6.2). A similar agreement is found once R_{CP+} and $\mathbf{A}_{CP+}$ are derived from the best fitted r_B^K, δ_B^K and γ by the CKMfitter group [9].[1]
- **Suppressed modes**: $R_{\mathbf{ADS}}$ **and** $\mathbf{A}_{\mathbf{ADS}}$
  The $R_{\mathbf{ADS}}$ and $\mathbf{A}_{\mathbf{ADS}}$ observables are plotted in Figs. 6.7 and 6.8 for $B \to DK$ and $B \to D\pi$ decays respectively. A good agreement is found for all the measured quantities and all uncertainties are found to be comparable to those of the averaged values of the existing measurements, computed without including this analysis.

[1] It should be noted that the global fit to r_B^K, δ_B^K and γ is dominated by the b-factory Dalitz analyses

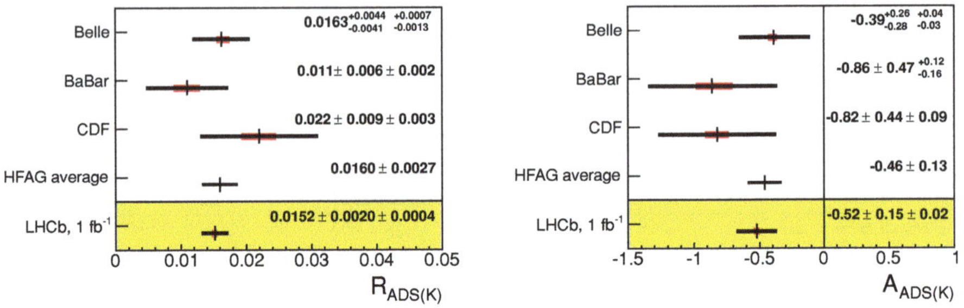

Fig. 6.7 Comparison plot of of the measured $R_{\mathbf{ADS}(K)}$ and $A_{\mathbf{ADS}(K)}$ observables

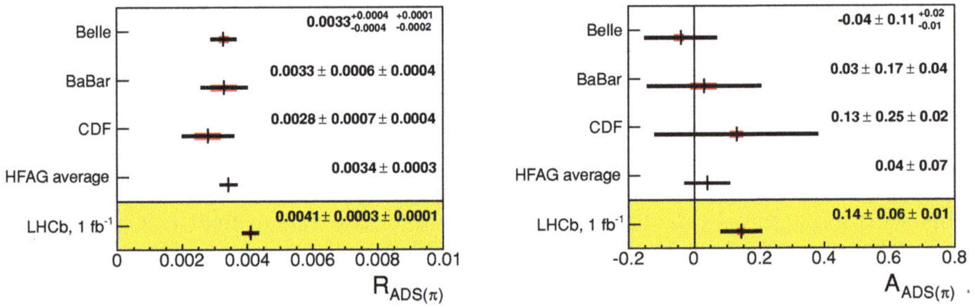

Fig. 6.8 Comparison plot of of the measured $R_{\mathbf{ADS}(\pi)}$ and $A_{\mathbf{ADS}(\pi)}$ observables

Interestingly, $A_{\mathbf{ADS}(\pi)}$ is consistently higher than the b-factories (although compatible with 0 at 2σ level).

6.6 A Preliminary Fit to the CKM Angle γ

As seen in Sect. 1.6.3, different quantities can be used to extract the CKM phase γ at tree level. The general idea is to measure as many γ-related quantities and extract all the r_B and δ_B hadronic parameters from data itself, where possible. Different final states can be used to gain information on γ, although different modes have different r_B and δ_B variables (which depend on the particular B and D decay considered) increasing the number of unknowns of the system. It should be noted that the number of equations scales approximately quadratically as $\sim(N_B \cdot N_D)$, where N_B and N_D are the numbers of different B and D modes considered in the scan. On the other hand, the number of unknowns (the hadronic parameters) only increases as $\sim(N_B + N_D)$, but most of the parameters are in common, including the weak phase γ. It is clear that, if enough quantities are measured at the same time, the system is in principle solvable and all unknowns can be extracted from the experiment. In LHCb, this simultaneous extraction would also benefit of a combined treatment of systematic uncertainties, which are expected to be correlated among the different modes considered. Furthermore, as the equations used are in general non-linear in the parameters, including trigonometric functions, the fit to γ is not trivial and might contain ambiguities.

6.6.1 Likelihood Scan Including Only $B \to [hh]_D K$ Observables

This section present a preliminary three-dimensional likelihood scan on the $(r_B^K, \delta_B^K, \gamma)$ parameters. The following *CP*-related quantities (see Sect. 1.6.3) are used in the scan:

$$A_{CP+} = \frac{2r_B^K \sin\delta_B^K \sin\gamma}{1 + {r_B^K}^2 + 2r_B^K \cos\delta_B^K \cos\gamma} \tag{6.8}$$

$$A_K^{K\pi} = \frac{2r_B^K r_D \sin(\delta_B^K - \delta_D)\sin\gamma}{1 + (r_B^K r_D)^2 + 2r_B^K r_D \cos(\delta_B^K - \delta_D)\cos\gamma} \tag{6.9}$$

$$R_{CP+} = 1 + {r_B^K}^2 + 2r_B^K \cos\delta_B^K \cos\gamma \tag{6.10}$$

$$R_K^- = \frac{{r_D}^2 + {r_B^K}^2 + 2r_B^K r_D \cos(\delta_B^K + \delta_D - \gamma)}{1 + {r_B^K}^2 {r_D}^2 + 2r_B^K r_D \cos(\delta_B^K - \delta_D - \gamma)} \tag{6.11}$$

$$R_K^+ = \frac{{r_D}^2 + {r_B^K}^2 + 2r_B^K r_D \cos(\delta_B^K + \delta_D + \gamma)}{1 + {r_B^K}^2 {r_D}^2 + 2r_B^K r_D \cos(\delta_B^K - \delta_D + \gamma)} \tag{6.12}$$

Only asymmetries and ratios deriving from $B \to DK$ decays, where most of the sensitivity to γ is expected, are included in this scan.[2] The r_D and δ_D parameters are measured with good accuracy in $D \to K\pi$ decays and are fixed to be [10]:

$$r_D = (6.16 \pm 0.15)\,\% \tag{6.13}$$

$$\delta_D = (202.4^{+9.7}_{-11.0})^\circ \tag{6.14}$$

The likelihood function is defined, with a multi-dimensional Gaussian approach, as:

$$L(r_B^K, \delta_B^K, \gamma) = \exp\left(-\frac{1}{2}\,[\vec{\mathbf{x}}_{\text{meas}} - \vec{\mathbf{x}}_{\text{true}}]^T\, \mathbf{V}_{\text{TOT}}^{-1}\, [\vec{\mathbf{x}}_{\text{meas}} - \vec{\mathbf{x}}_{\text{true}}]\right) \tag{6.15}$$

where $\vec{\mathbf{x}}_{\text{meas}}$ is the vector of the measured quantities, $\vec{\mathbf{x}}_{\text{true}}(r_B^K, \delta_B^K, \gamma)$ are the corresponding *true* values computed from the fundamental parameters $(r_B^K, \delta_B^K, \gamma)$ and the normalisation factor has been dropped. $\mathbf{V}_{\text{TOT}} = \mathbf{V}_{\text{stat}} + \mathbf{V}_{\text{sys}}$ is the covariance

---

[2] $B \to D\pi$ related quantities carry less sensitivity to γ due to the lower value of r_B^π and two additional hadronic parameters (r_B^π, δ_B^π) must be introduced.

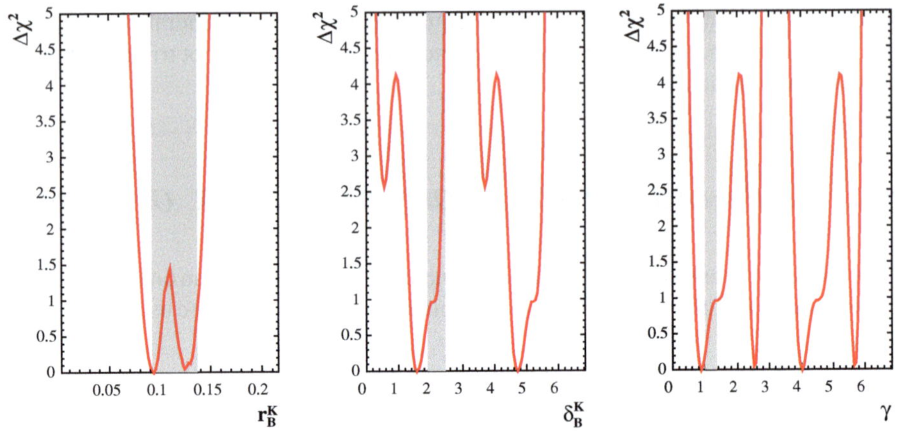

Fig. 6.9 1D projections of the (r_B^K,δ_B^K,γ) scan. Each curve is obtained minimising the $\Delta\chi^2$ function with respect to the other variables. The *shaded areas* indicate the current best estimated values [10]

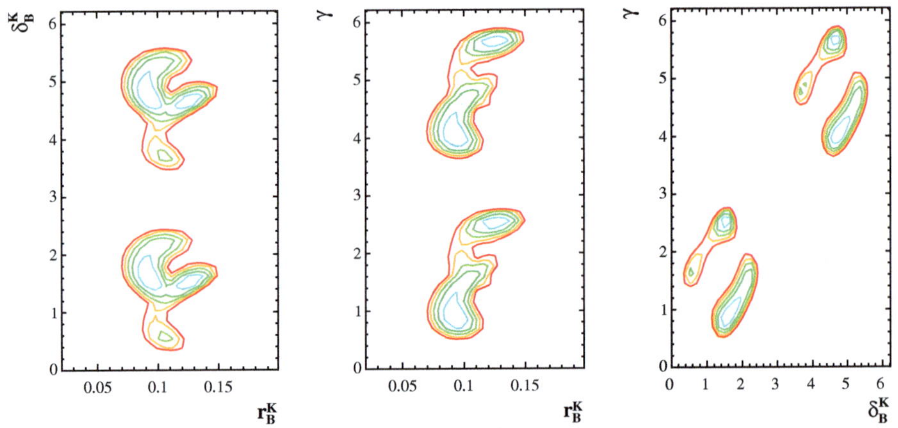

Fig. 6.10 2D projections of the (r_B^K,δ_B^K,γ) scan. Each 2D curve is obtained minimising the $\Delta\chi^2$ function with respect to the other variable. 1σ to 5σ contours are plotted

matrix, which takes into account any possible correlation among the observables.[3] It is convenient to plot the quantity (which provides a measure of the significance):

$$\Delta\chi^2(r_B^K, \delta_B^K, \gamma) = -2\ln L(r_B^K, \delta_B^K, \gamma) = [\vec{\mathbf{x}}_{\text{meas}} - \vec{\mathbf{x}}_{\text{true}}]^T\, \mathbf{V}_{\text{TOT}}^{-1}\, [\vec{\mathbf{x}}_{\text{meas}} - \vec{\mathbf{x}}_{\text{true}}] \tag{6.16}$$

(r_B^K,δ_B^K,γ) triplets are scanned on a (100, 100, 100) point grid. 1D and 2D projections are shown in Figs. 6.9 and 6.10 respectively. The CKM angle γ exhibits a four-fold ambiguity, which cannot be resolved using only the equations coming from the

[3] Only statistical correlations are considered, systematical errors are assumed Gaussian with no correlations.

ADS/GLW observables. However, it should be noted that γ is excluded from 0 with a significance $> 5\sigma$. $\gamma \neq 0$ and $\delta_B \neq 0$ are two requirements needed for direct *CP* violation (see Sect. 1.4.1).

6.6.2 Likelihood Scan Including the GGSZ Analysis

Several different modes can be exploited to extract the CKM phase γ. $B \to DK$ decays can be reconstructed where the D meson decays in a three-body, self-conjugate final state, e.g. $D \to K_S^0\pi^+\pi^-$ or $D \to K_S^0K^+K^-$. The use of $B^\pm \to [K_S^0h^+h^-]_D K^\pm$ modes is usually referred to as GGSZ (or *Dalitz* method) from the names of the proponents [11]. The GGSZ method is not discussed here in detail, however its basic idea is to exploit the two-dimensional mirror-symmetry of the $D \to K_S^0h^+h^-$ Dalitz plot, where different interference patterns can occur, in order to measure the *CP* observables $x_\pm$ and $y_\pm$, which are related to r_B^K,δ_B^K and γ via:

$$x_\pm = r_B^K \cos(\delta_B^K \pm \gamma) \tag{6.17}$$

$$y_\pm = r_B^K \sin(\delta_B^K \pm \gamma) \tag{6.18}$$

These four variables can thus be included in the scan of (r_B^K,δ_B^K,γ), providing extra constraints on the fundamental parameters. Thc LHCb experiment has recently measured [12]:

$$x_- = (\ \ 0.0 \pm 4.3 \pm 1.5 \pm 0.6) \times 10^{-2} \tag{6.19}$$

$$y_- = (\ \ 2.7 \pm 5.2 \pm 0.8 \pm 2.3) \times 10^{-2} \tag{6.20}$$

$$x_+ = (-10.3 \pm 4.5 \pm 1.8 \pm 1.4) \times 10^{-2} \tag{6.21}$$

$$y_+ = (-0.9 \pm 3.7 \pm 0.8 \pm 3.0) \times 10^{-2} \tag{6.22}$$

where the first uncertainty is statistical, and the second and the third systematical. Once Eqs. 6.17 and 6.18 are included in the construction of $\Delta\chi^2(r_B^K, \delta_B^K, \gamma)$ (see Eq. 6.16), their structure allows the scan to resolve the intrinsic ambiguities obtained when only the $B \to [hh]_D K$ observables are used (modulo π).

6.6.3 Including the Systematical Uncertainties on r_D and δ_D

The systematical uncertainties on the measured *CP*-related observables are included in the total covariance matrix, which is defined as $\mathbf{V}_{\text{TOT}} = \mathbf{V}_{\text{stat}} + \mathbf{V}_{\text{sys}}$. However, the hadronic parameters r_D and δ_D have been fixed so far to their measured central values presented in Eqs. 6.13 and 6.14 and a systematical uncertainty must be assessed for

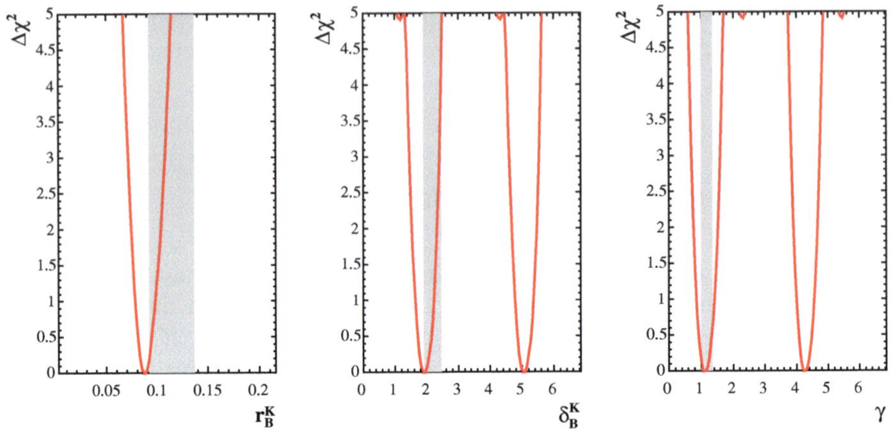

Fig. 6.11 1D projections of the (r_B^K,δ_B^K,γ) scan including the GGSZ analysis and the systematical uncertainties on r_D and δ_D. Each curve is obtained minimising the $\Delta\chi^2$ function with respect to the other variables. The *shaded areas* indicate the current best estimated values [10]

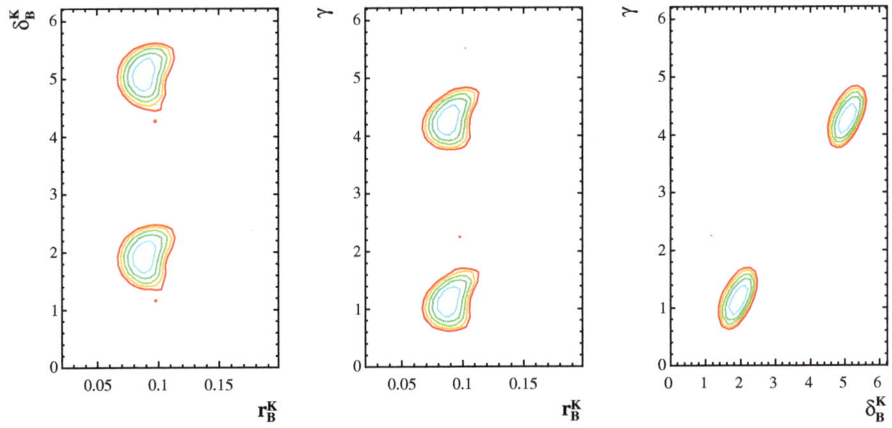

Fig. 6.12 2D projections of the (r_B^K,δ_B^K,γ) scan including the GGSZ analysis and the systematical uncertainties on r_D and δ_D. Each 2D curve is obtained minimising the $\Delta\chi^2$ function with respect to the other variable. 1σ to 5σ contours are plotted

this assumption. Eq. 6.15 is generalised as follows:

$$L_{\text{TOT}}(r_B^K, \delta_B^K, \gamma\,|r_D, \delta_D) = L(r_B^K, \delta_B^K, \gamma, r_D, \delta_D) \cdot P(r_D, \delta_D) \qquad (6.23)$$

where $P(r_D, \delta_D)$ represents the probability density function of the r_D and δ_D parameters. $P(r_D, \delta_D)$ should be centered on the measured central values, with some smearing depending on the quoted experimental error. An obvious choice is to assume uncorrelated, Gaussian errors, therefore:

Table 6.5 Contour levels of the 1D scan of the physical parameters r_B^K, δ_B^K and γ

Parameter	Central $\pm 1\sigma$	Central $\pm 2\sigma$	Central $\pm 3\sigma$
r_B^K	$0.087^{+0.011}_{-0.009}$	$0.087^{+0.017}_{-0.012}$	$0.087^{+0.021}_{-0.015}$
δ_B^K [°]	$111.6^{+16.2}_{-17.5}$	$111.6^{+21.7}_{-23.5}$	$111.6^{+25.8}_{-27.7}$
γ [°]	$64.8^{+16.6}_{-15.6}$	$64.8^{+22.7}_{-20.9}$	$64.8^{+26.8}_{-24.5}$

$$L_{\text{TOT}}(r_B^K, \delta_B^K, \gamma \,|r_D, \delta_D) = L(r_B^K, \delta_B^K, \gamma, r_D, \delta_D) \cdot \exp\left(-\frac{(r_D - \mu_{r_D})^2}{2\sigma_{r_D}^2}\right) \cdot \exp\left(-\frac{(\delta_D - \mu_{\delta_D})^2}{2\sigma_{\delta_D}^2}\right) \tag{6.24}$$

The minimisation is performed on the quantity:

$$\begin{aligned}\Delta\chi^2_{\text{TOT}}(r_B^K, \delta_B^K, \gamma \,|r_D, \delta_D) &= -2\ln L_{\text{TOT}}(r_B^K, \delta_B^K, \gamma \,|r_D, \delta_D) \\ &= \Delta\chi^2(r_B^K, \delta_B^K, \gamma \,|r_D, \delta_D) + \frac{(r_D - \mu_{r_D})^2}{2\sigma_{r_D}^2} + \frac{(\delta_D - \mu_{\delta_D})^2}{2\sigma_{\delta_D}^2}\end{aligned} \tag{6.25}$$

where $\Delta\chi^2$ is obtained as in Eq. 6.16. The scan is performed in the same way as Sect. 6.6.1 and Sect. 6.6.2 but it is now extended to r_D and δ_D as well. The corresponding 3D $(r_B^K, \delta_B^K, \gamma)$ projection is obtained minimising $\Delta\chi^2_{\text{TOT}}(r_B^K, \delta_B^K, \gamma|r_D, \delta_D)$ on the (r_D, δ_D) plane. The effect of the inclusion of the GGSZ analysis and the systematical smearing is shown in Figs. 6.11 and 6.12. The 1σ, 2σ and 3σ contour levels of Fig. 6.11 are reported in Table 6.5.

References

1. LHCb collaboration, Evidence for the suppressed decay $B^\pm \to (K^\mp\pi^\pm)_D K^\pm$, LHCb-CONF-2011-044, LHCb-ANA-2011-046 (2011)
2. K. Nakamura et al., (Particle Data Group) Review of particle physics. J. Phys. G37, 075021 (2010)
3. BaBar collaboration, P. del Amo Sanchez et al., Measurement of CP observables in $B^\pm \to D_{CP}K^\pm$ decays and constraints on the CKM angle γ. Phys. Rev. D82 072004, arXiv:1007.0504 (2010)
4. Belle collaboration, K. Abe et al., Study of $B^\pm \to D_{CP}K^\pm$ and $D^*_{CP}K^\pm$ decays. Phys. Rev. D73 051106, arXiv:hep-ex/0601032 (2006)
5. Belle collaboration, Lepton photon 2011 preliminary (2011)
6. CDF collaboration, T. Aaltonen et al., Measurements of branching fraction ratios and CP asymmetries in $B^\pm \to D_{CP}K^\pm$ decays in hadron collisions. Phys. Rev. D81 031105, arXiv:0911.0425 (2010)

7. BaBar collaboration, P. del Amo Sanchez et al., Evidence for direct CP violation in the measurement of the CKM angle γ with $B^{\mp} \to D^{(*)}K^{(*)\mp}$ decays. Phys. Rev. Lett. 105, 121801, arXiv:1005.1096 (2010)
8. Belle collaboration, A. Poluektov et al., Evidence for direct CP violation in the decay $B^{\pm} \to D^{(*)}K^{\pm}$, $D \to K_S^0\pi^+\pi^-$ and measurement of the CKM phase ϕ_3. Phys. Rev. D81 112002, arXiv:1003.3360 (2010)
9. CKMfitter group, J. Charles et al., Predictions of selected flavor observables within the standard model. Phys. Rev. D 84 033005 (2011)
10. Heavy Flavor Averaging Group, Y Amhis et al., Averages of b-hadron, c-hadron, and tau-lepton properties as of early 2012, arXiv:1207.1158 (2012)
11. Giri, Anjan and Grossman, Yuval and Soffer, Abner and Zupan, Jure, Determining γ using $B^{\pm} \to DK^{\pm}$ with multibody D decays. Phys. Rev. D 68 (2003) 054018
12. D. Johnson, S. Malde, G. Wilkinson, A model-independent analysis of $D \to K_s^0h^+h^-$ Dalitz plots from $B^{\pm} \to Dh^{\pm}$ decays. Technical representation LHCb-PAPER-2012-027, CERN, Geneva (2012)

Chapter 7
Conclusions

A precise measurement of the CKM angle γ is a core goal for the LHCb Physics program. In this thesis, an analysis of CP violation in $B^{\pm} \rightarrow DK^{\pm}$ and $B^{\pm} \rightarrow D\pi^{\pm}$ decays is presented. D mesons are reconstructed in two-body charged final states: $K^{\pm}\pi^{\mp}$, $K^{+}K^{-}$, $\pi^{+}\pi^{-}$ and $\pi^{\pm}K^{\mp}$. CP-related observables are extracted via a simultaneous fit to the invariant B mass distributions of the modes considered. The full 2011 LHCb dataset of $1.0\,\text{fb}^{-1}$ is used for the analysis and results have been published in [1]. All measurements are compatible with the current world averages and they are the most competitive:

$$\begin{aligned}
R_{CP+} &= 1.007 \pm 0.038 \pm 0.012 \\
A_{CP+} &= 0.145 \pm 0.032 \pm 0.010 \\
R_{\text{ADS}(K)} &= 0.0152 \pm 0.0020 \pm 0.0004 \\
A_{\text{ADS}(K)} &= -0.52 \pm 0.15 \pm 0.02 \\
R_{\text{ADS}(\pi)} &= 0.00410 \pm 0.00025 \pm 0.00005 \\
A_{\text{ADS}(\pi)} &= 0.143 \pm 0.062 \pm 0.011
\end{aligned}$$

The suppressed $B^{\pm} \rightarrow DK^{\pm}$ mode is observed for the first time with $\approx 10\sigma$ significance. This mode displays evidence (4.0σ) of a large negative asymmetry. The suppressed $B^{\pm} \rightarrow D\pi^{\pm}$ mode also exhibits hints of a positive asymmetry (2.4σ significance), which could be used to improve the sensitivity on γ. Once all measurements are combined, direct CP violation is established in $B^{\pm}$ decays with a total combined significance of 5.8σ. The results presented in this thesis are used as inputs to a preliminary fit to the CKM phase γ. γ is excluded from 0 with a significance $>5\sigma$, using only information coming from the analysis presented.
The analysis will be updated using the dataset collected in 2012, with a total integrated luminosity summing to over 2.5fb^{-1}. The phase γ will be extracted with an expected precision, i.e. combining with other modes, of about $10°$ by end 2013, and of 5–7 ° by the the time LHCb data taking stops in 2017.

P. Gandini, *Observation of CP Violation in $B^{\pm} \rightarrow DK^{\pm}$ Decays*,
Springer Theses, DOI: 10.1007/978-3-319-01029-8_7,

Reference

1. The LHCb Collaboration, R. Aaij et al. Observation of CP violation in $B^+ \rightarrow DK^+$ decays. Phys. Lett. **B712**, 203 (2012, arXiv:1203.3662)

Appendix A
Distributions of the BDT Input Variables

This section shows the distribution of the signal/background distributions of the BDT input variables. When variables present long tails in their distribution, the logarithm of the variable is preferred and used in the training and evaluation procedure.

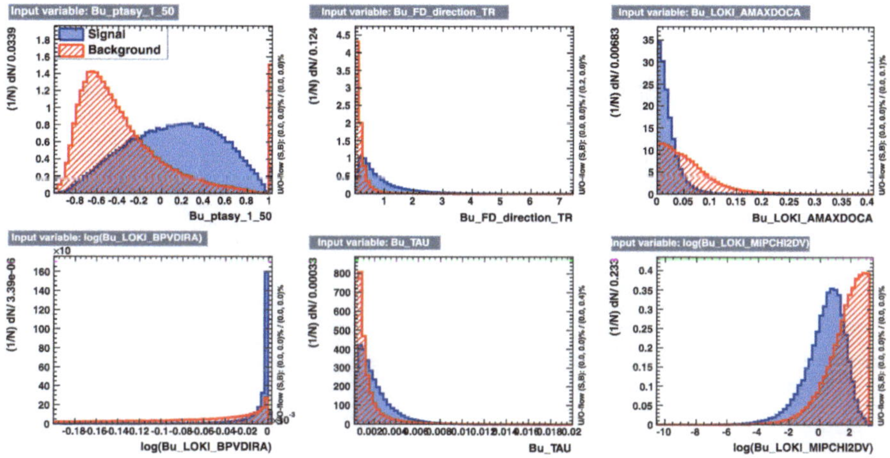

P. Gandini, *Observation of CP Violation in $B^{\pm} \to DK^{\pm}$ Decays*,
Springer Theses, DOI: 10.1007/978-3-319-01029-8,

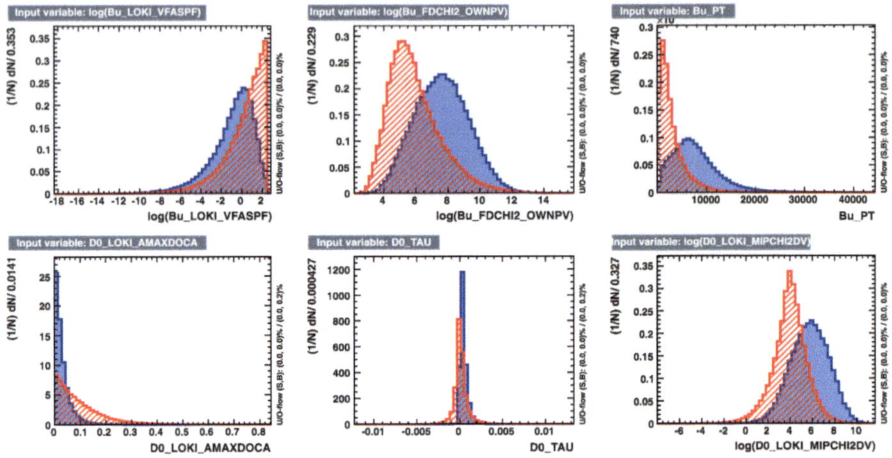
Input variable: log(Bu_LOKI_VFASPF)
log(Bu_LOKI_VFASPF)
Input variable: log(Bu_FDCHI2_OWNPV)
log(Bu_FDCHI2_OWNPV)
Input variable: Bu_PT
Bu_PT
Input variable: D0_LOKI_AMAXDOCA
D0_LOKI_AMAXDOCA
Input variable: D0_TAU
D0_TAU
Input variable: log(D0_LOKI_MIPCHI2DV)
log(D0_LOKI_MIPCHI2DV)

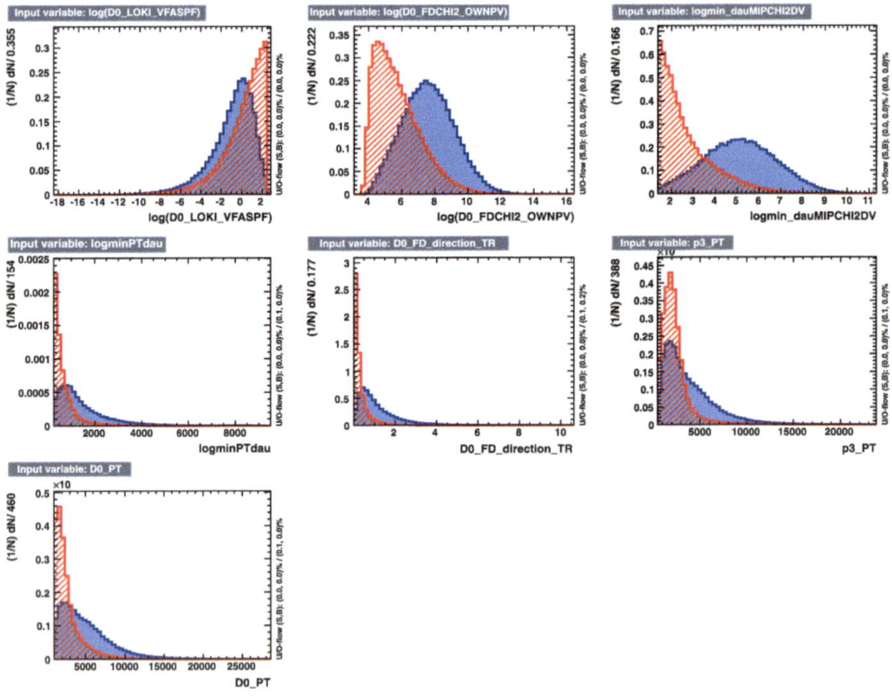
Input variable: log(D0_LOKI_VFASPF)
log(D0_LOKI_VFASPF)
Input variable: log(D0_FDCHI2_OWNPV)
log(D0_FDCHI2_OWNPV)
Input variable: logmin_dauMIPCHI2DV
logmin_dauMIPCHI2DV
Input variable: logminPTdau
logminPTdau
Input variable: D0_FD_direction_TR
D0_FD_direction_TR
Input variable: p3_PT
p3_PT
Input variable: D0_PT
D0_PT

Appendix B
Validation of the A_{p_T} Variable

Since MC simulated events are used in the BDT training, as described in Sect. 3.5.1, a good MC/data agreement is required in the description of the full event inside the LHCb detector. For this purpose, MC simulated $B \to [K\pi]_D\pi$ signal events have been compared to a data sample of $B \to [K\pi]_D\pi$ candidates taken from the 2010 $35.7\mathrm{pb}^{-1}$ sample, the same used in the training, required to be inside both the B and D tight mass windows. The samples are superimposed for different values of the θ_cone aperture angle. A good agreement in the $A_{p_\mathrm{T}} \in [-1, +1)$ range is found, although less totally isolated events are found in data with respect to the MC simulated sample, as expected from the LHCb central pp event simulation production.

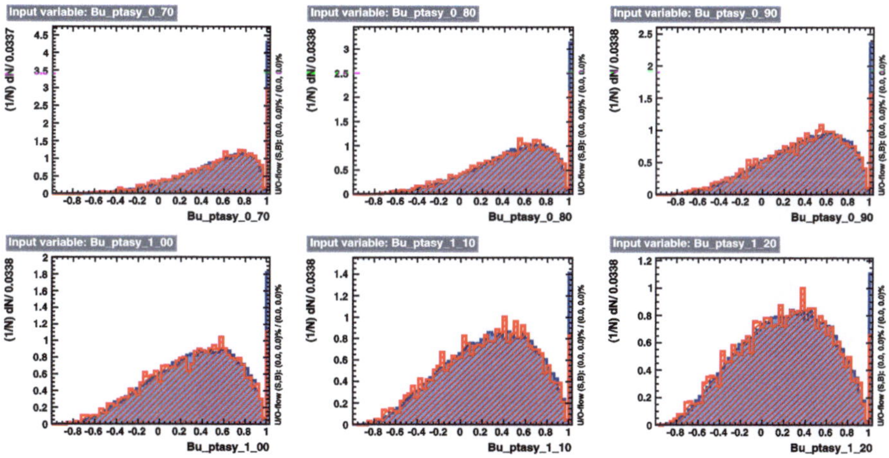

P. Gandini, *Observation of CP Violation in $B^\pm \to DK^\pm$ Decays*,
Springer Theses, DOI: 10.1007/978-3-319-01029-8,

Appendix C
Histograms of Systematic Errors

In order to evaluate the effect of each systematic uncertainty, the fit is rerun many times in his final configuration, varying each of the fixed constraints (the fixed parameters introduced in the fitter) according to a Gaussian distribution. For every systematic source, a histogram is filled with as many entries as the repeated fits. In this way, it is possible to evaluate the effect of each assumption on the physics observables separately and judge any possible non-Gaussian behaviour.

The distribution of $R^{K\pi}_{K/\pi}$ results under systematic variations

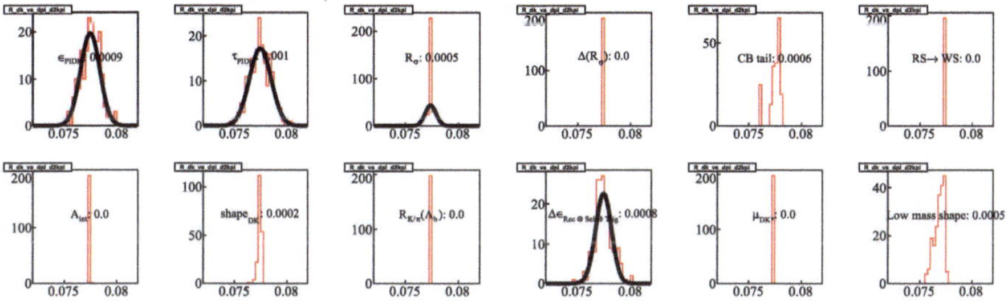

P. Gandini, *Observation of CP Violation in $B^{\pm} \rightarrow DK^{\pm}$ Decays*,
Springer Theses, DOI: 10.1007/978-3-319-01029-8,

The distribution of $R^{KK}_{K/\pi}$ results under systematic variations

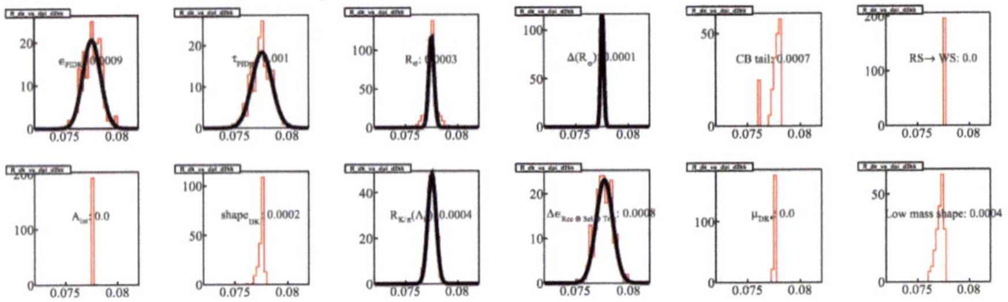

The distribution of $R^{\pi\pi}_{K/\pi}$ results under systematic variations

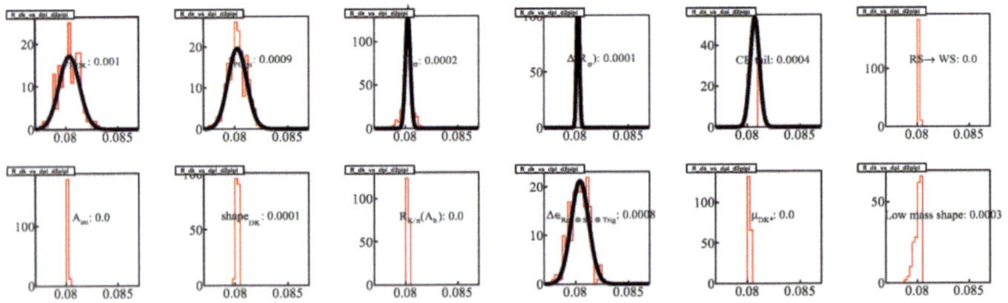

The distribution of $A^{K\pi}_{\pi}$ results under systematic variations

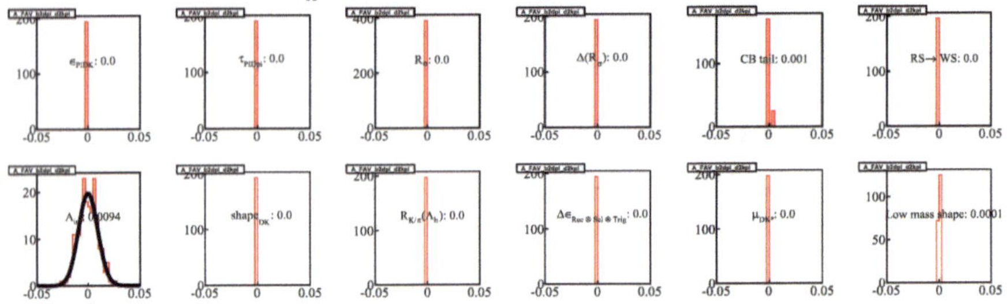

The distribution of $A^{K\pi}_{K}$ results under systematic variations

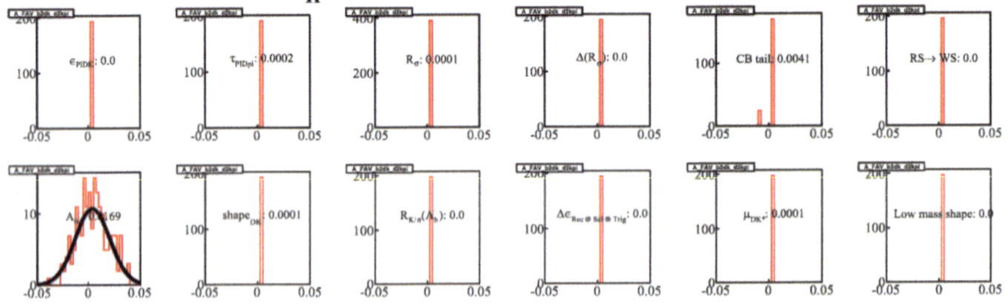

The distribution of A_{π}^{KK} results under systematic variations

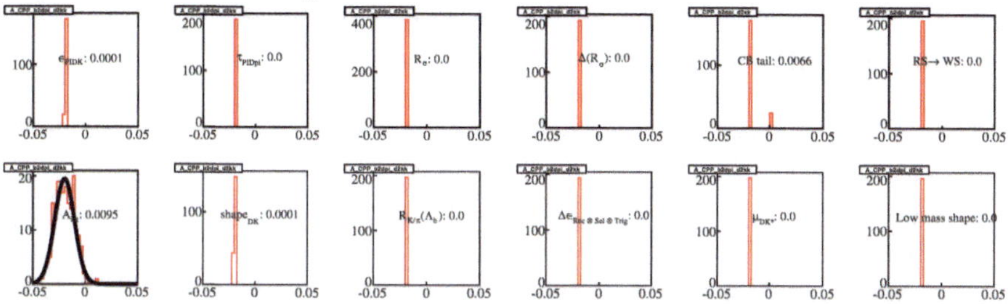

The distribution of $A_{\pi}^{\pi\pi}$ results under systematic variations

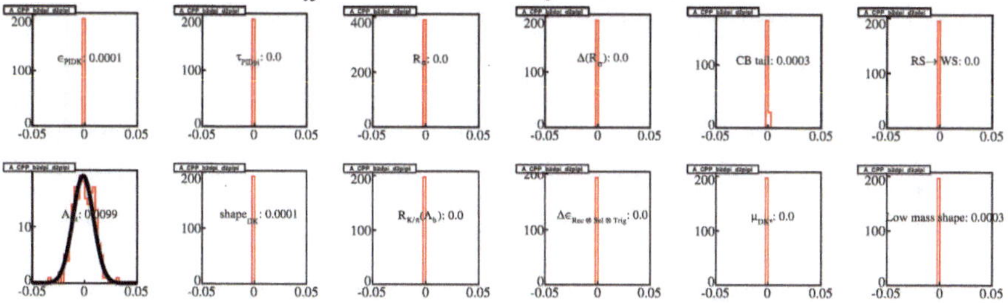

The distribution of A_{K}^{KK} results under systematic variations

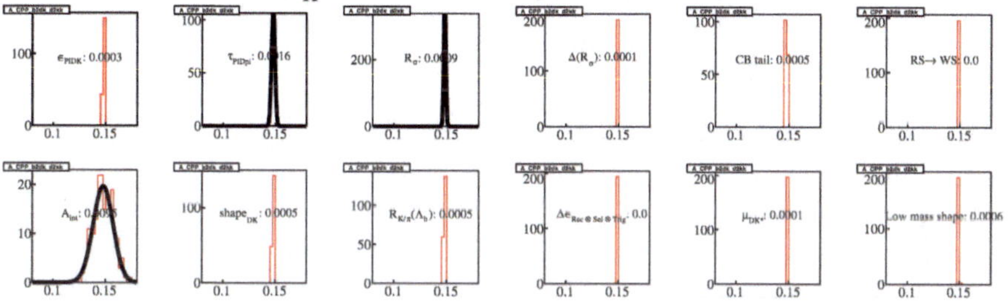

The distribution of $A_{K}^{\pi\pi}$ results under systematic variations

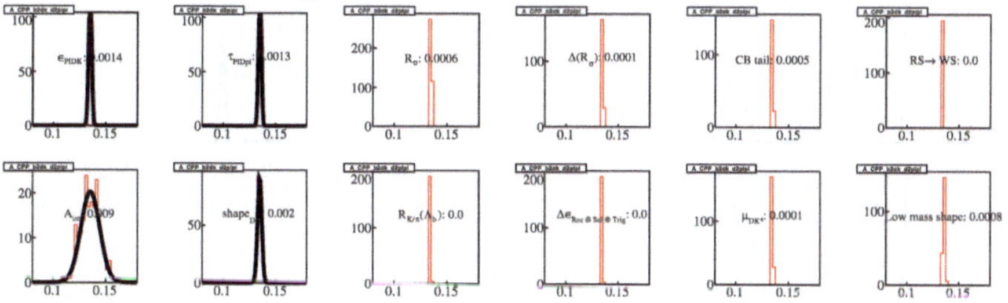

The distribution of R_{π}^{-} results under systematic variations

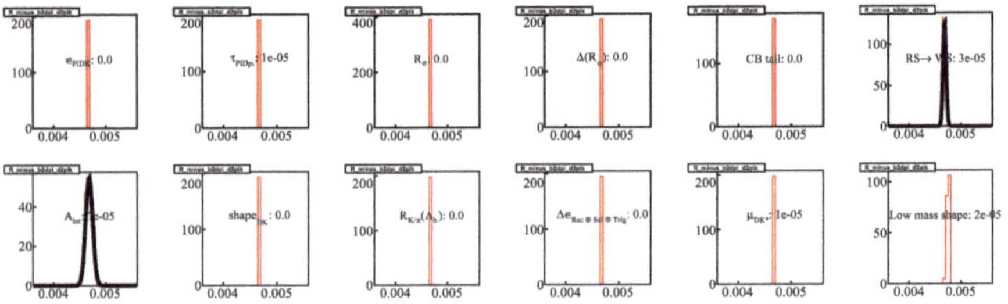

The distribution of R_{π}^{+} results under systematic variations

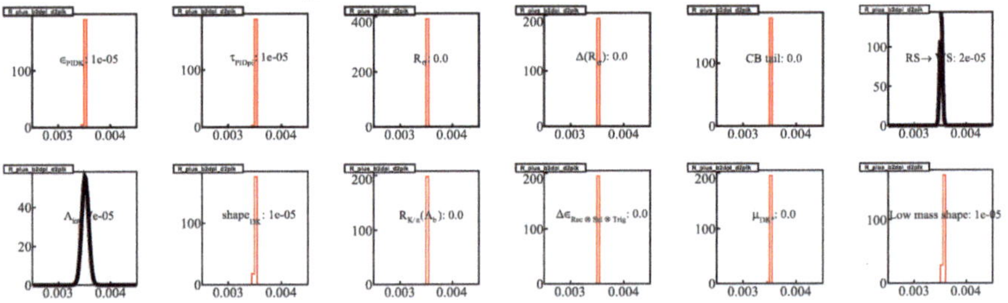

The distribution of R_{K}^{-} results under systematic variations

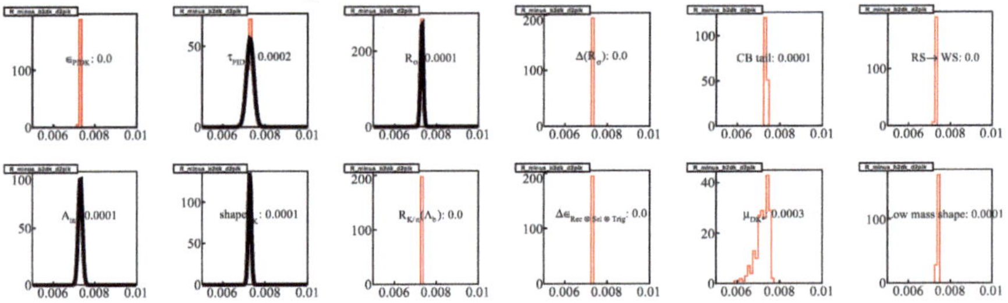

The distribution of R_{K}^{+} results under systematic variations

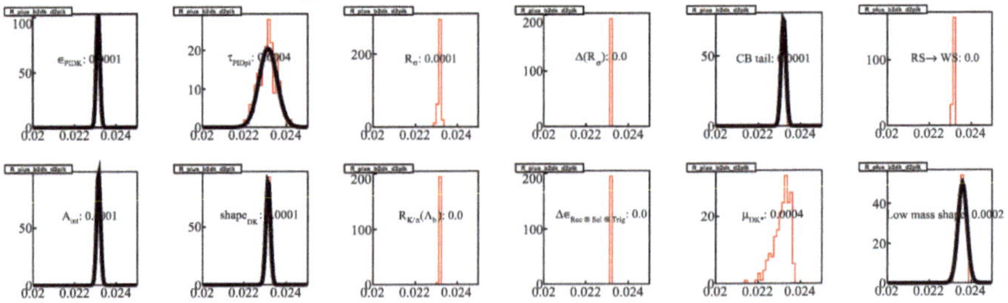

Appendix D
Curriculum Vitae

Paolo Gandini is a particle physicist mainly interested in Flavour Physics. Currently member of the LHCb collaboration at CERN, he completed a D.Phil in Particle Physics at the University of Oxford with a thesis on *CP* violation studies in charged *B* decays. He is currently based at CERN as a postdoctoral researcher for Syracuse University, NY. Member of the author list of the LHCb Collaboration from September 2009).

Education

Jan 2013	**Postdoctoral researcher for Syracuse University, NY**
Oct 2012	**D.Phil. in Experimental Particle Physics** Balliol College, University of Oxford Thesis: "Observation of CP Violation in $B^{\pm} \rightarrow DK^{\pm}$ decays (LHCb)" Supervisor: Dr Malcolm John, University of Oxford
Oct 2008	**Master of Science in Physics** Università degli Studi di Milano, Milan—*Grade 110/110 cum Laude* Thesis: "Branching Fraction measurement of the B^0 meson decay to $a_1^+(1260)\, a_1^-(1260)$ (BaBar)" Supervisor: Prof. Fernando Palombo (University of Milan, INFN Milan)
Oct 2006	**Bachelor of Science in Physics** Università degli Studi di Milano, Milan—*Grade 110/110 cum Laude* Thesis: "Research of SUSY particles in di-lepton decays with ATLAS detector at LHC" Supervisor: Prof. F. Ragusa (University of Milan, INFN Milan)

Awards

2009–2012	JT Hamilton Scholarship in Physics, Balliol College, Oxford

P. Gandini, *Observation of CP Violation in $B^{\pm} \rightarrow DK^{\pm}$ Decays*,
Springer Theses, DOI: 10.1007/978-3-319-01029-8,

MIX
Papier aus verantwortungsvollen Quellen
Paper from responsible sources
FSC® C105338

If you have any concerns about our products,
you can contact us on
ProductSafety@springernature.com

In case Publisher is established outside the EU,
the EU authorized representative is:
Springer Nature Customer Service Center GmbH
Europaplatz 3, 69115 Heidelberg, Germany

Printed by Libri Plureos GmbH
in Hamburg, Germany